Water Science Reviews 4

Water Science Reviews 4

Hydration Phenomena in Colloidal Systems

EDITED BY

FELIX FRANKS

CAMBRIDGE UNIVERSITY PRESS

CAMBRIDGE

NEW YORK PORT CHESTER

MELBOURNE SYDNEY

CAMBRIDGE UNIVERSITY PRESS
Cambridge, New York, Melbourne, Madrid, Cape Town, Singapore, São Paulo, Delhi

Cambridge University Press
The Edinburgh Building, Cambridge CB2 8RU, UK

Published in the United States of America by Cambridge University Press, New York

www.cambridge.org
Information on this title: www.cambridge.org/9780521365789

First published 1989
This digitally printed version 2008

A catalogue record for this publication is available from the British Library

Library of Congress Cataloguing in Publication data

Water science reviews. – 1- – Cambridge; New York: Cambridge
University Press, 1985-
v.: ill.; 24 cm.
Annual.
Editor: 1985- F. Franks.
1. Water chemistry—Periodicals. 2. Water—Periodicals. 3. Water Dynamics—Periodicals.
I. Franks, Felix.
GB855.W38 546′.22′05—dc19 86-643278
AACR 2 MARC-S
Library of Congress 8705

ISBN 978-0-521-36578-9 hardback
ISBN 978-0-521-10025-0 paperback

Water Science Reviews 4

Hydration Phenomena in Colloidal Systems

EDITED BY

FELIX FRANKS

CAMBRIDGE UNIVERSITY PRESS

CAMBRIDGE

NEW YORK PORT CHESTER

MELBOURNE SYDNEY

CAMBRIDGE UNIVERSITY PRESS
Cambridge, New York, Melbourne, Madrid, Cape Town, Singapore, São Paulo, Delhi

Cambridge University Press
The Edinburgh Building, Cambridge CB2 8RU, UK

Published in the United States of America by Cambridge University Press, New York

www.cambridge.org
Information on this title: www.cambridge.org/9780521365789

First published 1989
This digitally printed version 2008

A catalogue record for this publication is available from the British Library

Library of Congress Cataloguing in Publication data

Water science reviews. – 1- – Cambridge; New York: Cambridge
University Press, 1985-
v.: ill.; 24 cm.
Annual.
Editor: 1985- F. Franks.
1. Water chemistry—Periodicals. 2. Water—Periodicals. 3. Water Dynamics—Periodicals.
I. Franks, Felix.
GB855.W38 546′.22′05—dc19 86-643278
AACR 2 MARC-S
Library of Congress 8705

ISBN 978-0-521-36578-9 hardback
ISBN 978-0-521-10025-0 paperback

Contents

A reappraisal of the role of water in promoting amphiphilic assembly and structure

D. FENNELL EVANS AND DAVID D. MILLER

Department of Chemical Engineering and Materials Science, University of Minnesota, Minneapolis MN 55455, USA

1. Introduction

Since the evolution of life began, nature has utilized amphiphilic molecules to form self-organizing structures possessing oil-like interiors and large interfacial areas. Such amphiphilic molecules self-assemble in water to form a variety of microstructures including micelles, vesicles, liposomes, microtubules and bilayers. Phospholipid bilayers act as controlled-access barriers isolating each cell from its environment. Through subtle changes in pH, temperature, ionic strength and counterion, living systems are able to transform continuously from one microstructure to another in response to physiological need. Examples are pinocytosis, endocytosis and phagocytosis. Such transformations guide virtually all biological processes.

Man's exploration of amphiphilic systems as cleaning agents and as decorative and protective coatings dates back to the earliest development of technology. Their utilization in 'high tech' applications, such as vesicular carriers for controlled drug delivery or as microdomains for controlled synthesis, depends upon the ability to set amphiphilic structure, maintain its integrity under adverse conditions and then transform it at the end of the process.

An understanding of biological processes at a level of sophistication that goes beyond stoichiometric biochemistry and the ability to utilize amphiphilic microstructures in practical applications requires an understanding of the intricate interactions between water and amphiphilic molecules. This is a complex issue. Self-assembly is a facile physicochemical process in which amphiphilic molecules are physically, not chemically, associated. Their microstructures can transform in response to small perturbations in their environment. In aqueous solutions these processes perturb water in ways that make it difficult to disentangle cause and effect.

For this reason, it is useful to step back and look at self-assembly in a more general sense. In the first part of this review, we examine aggregation phenomena in other solvents and then compare it to that observed in water. We next consider how solvent–amphiphile determine size, shape, reactivity and stability. These characteristics can be described conveniently in terms of intra- and intermolecular interactions. The former are illustrated by temperature effects and simple variations in counterion and are described in the second part of this review. In the final section, we discuss the myriad of forces that mediate the interaction between these colloidal entities.

The vast literature on amphiphilic self-assembly cannot be reviewed in a comprehensive way in this brief presentation. We will focus on recent studies of the alkyltrimethylammonium and dialkyldimethylammonium surfactants since these compounds illustrate many of the key points involved in understanding the self-organizing properties of amphiphilic molecules in aqueous solution.

2. Solvent–amphiphilic interactions that drive self-assembly

2.1. *General comments*

What solvent properties are necessary to promote the self-assembly of amphiphilic molecules? An obvious requirement is high solvent polarity. The microphasic separation that constitutes aggregation is a consequence of the low solubility of amphiphilic hydrocarbon chains. A useful measure of solvent polarity or cohesive energy density is the Gordon parameter [1] defined as $\gamma/V^{\frac{1}{3}}$ where γ is the surface tension and V is the molar volume. This quantity is analogous to the Hildebrand solubility parameter, but is more useful because it permits fused salts, such as ethylammonium nitrate (EAN), with vanishingly low vapor pressures to be considered. These ideas are summarized in figure 1 where the free energy of transfer of various gases from the vapor phase to selected solvents is displayed. Note that there is a roughly linear relationship between ΔG and $\gamma/V^{\frac{1}{3}}$ in all solvents except water. We will return to this point later.

Micelle formation has been documented in water, hydrazine,[2,3] ethylammonium nitrate,[4,5] formamide [6] and ethylene glycol [6] in which $\gamma/V^{\frac{1}{3}}$ decreases from 27 to 13 dyne cm^{-2}. As $\gamma/V^{\frac{1}{3}}$ decreases, the driving force for aggregation decreases causing the critical micelle concentrations (CMCs) to increase and become less sharp. Below a value of 13, no evidence of amphiphilic aggregation has been detected.

All of the solvents listed above possess the ability to form on the average three or four hydrogen bonds per solvent molecule and thus the ability to form a three-dimensional structure. Whether multiple hydrogen bonding is a prerequisite for amphiphilic aggregation has remained an unanswered question since all of the previous aprotic solvents investigated have Gordon parameters which are less than 13. Published reports of micelle formation in such solvents as dimethylsulfoxide (DMSO) have not been substantiated.

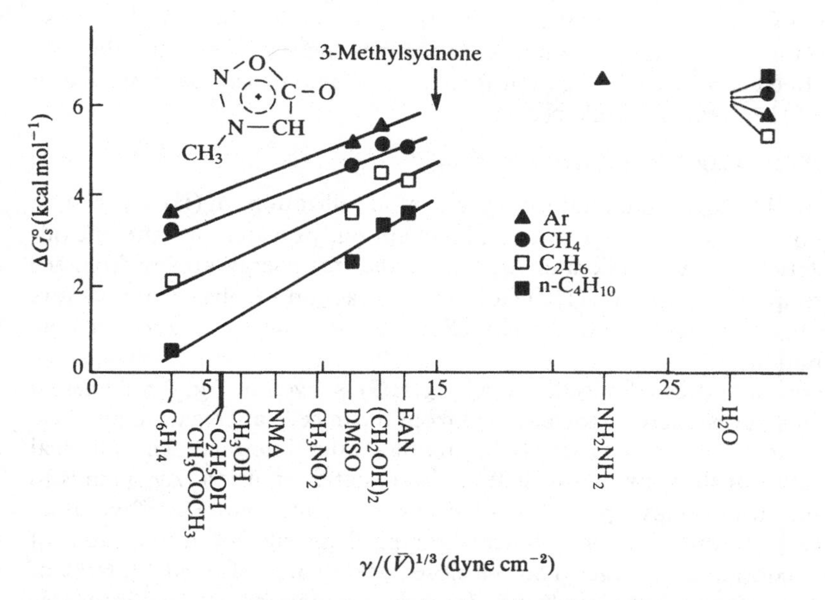

Figure 1. Free energy of transfer of argon, methane, ethane and n-butane from the gas phase to several liquids. The liquids are plotted according to their value of $\gamma/V^{\frac{1}{3}}$ where γ is the surface tension and V is the molar volume. This (Gordon) parameter is a measure of a liquid's 'cohesiveness' and liquids with Gordon parameters above about 13 tend to promote aggregation of amphiphilic molecules. One exception is the aprotic 3-methylsydnone where $\gamma/V^{\frac{1}{3}} \approx 15$, but where no aggregation phenomena have been detected (see text).

The aprotic solvent, 3-methylsydnone, possesses a high cohesive energy density ($\gamma/V^{\frac{1}{3}} \approx 15$), a large dielectric constant ($\epsilon = 144$ at 40 °C) and a large dipole moment ($\mu = 7.3$).[7] Since this solvent falls between hydrazine and EAN, one might expect it to be a good candidate for promoting aggregative phenomena. We have measured conductance and surface tension and carried out polarizing microscopy and diffusive interfacial transport (DIT) experiments [8] in this solvent on a variety of single- and double-chained amphiphiles.[9] We found no evidence for the amphiphilic self-assembly. Thus it would appear that another prerequisite for amphiphilic self-assembly is a solvent possessing the ability to form a three-dimensional structure via multiple hydrogen bonding.

In order to understand amphiphilic aggregation, it is useful to focus on the formation of spherical micelles. They have well-defined CMCs that are amenable to theoretical analysis. In addition, such aggregates are monodisperse, and interparticle interactions are at a minimum at the CMC. A well-defined CMC is the consequence of two competing factors in the formation of micelles. The first involves the transfer to the hydrocarbon chains out of water and into the oil-like interior of the micelles and drives micellization. The second arises from the repulsion between headgroups as they move into close proximity at the micellar surface and oppose micellization. This competition also determines the size of the micelle.

The free energy of micellization for ionic surfactants can be expressed in terms of the dressed micelle model as

$$RT \ln X_{CMC} = \Delta g(\text{HP}) + \Delta g(\text{HG}) \tag{1}$$

where $RT \ln X_{CMC}$ is the total free energy of micellization, $\Delta g(\text{HP})$ is the free energy of transferring the hydrocarbon chain out of water and into the oil-like interior of the micelle and $\Delta g(\text{HG})$ is the free energy arising from the headgroup interaction.[10,11] Headgroup interaction involves many factors including electrostatic and dipolar effects as well as steric and entropic contributions.

In terms of a dressed micelle model, $\Delta g(\text{HG})$ is given by $\Delta g_{el} + \gamma a$ where a is the area per headgroup (determined from the micelle aggregation number) and γ is the surface tension at the micelle–water interface. The physical significance of these two terms is that electrostatic repulsion (Δg_{el}) tends to push the headgroups apart, but in doing so creates more (unfavorable) hydrocarbon–water contacts. Since the typical micelle headgroup area of 60 Å^2 is considerably larger than the close packing area of ~ 30 Å^2, 60% of the surface of the micelle involves hydrocarbon–water contact. Δg_{el} is the free energy of assembling a double layer and is obtained (with a knowledge of the Debye length, κ^{-1}, and a) from solutions of the nonlinear Poisson–Boltzmann equation. Since for the optimal micelle, $\partial \Delta g_{el}/\partial a + \gamma = 0$, we can also calculate γ as $-\partial \Delta g_{el}/\partial a$. Therefore, $\Delta g(\text{HP})$ is also determined. As a check on these calculations, we can estimate $\Delta g(\text{HP})$ by summing the contributions from each methylene group (-720 cal mol^{-1}) and the methyl group (-2300 cal mol^{-1}).[12] Such calculations confirm the validity of the dressed micelle model. It should be mentioned that the dressed micelle model clarifies the physics underlying the equilibrium model of micellization.[11]

2.2 *Nonaqueous solvents*

We consider first micellization in hydrazine.[2] The reason for this choice is apparent from the data given in table 1. Hydrazine and water have astonishingly similar physical properties and both are capable of forming, on average, four hydrogen bonds per molecule. In fact, they differ in only those properties that have been associated with the unique structural properties of water such as heat capacity, density maximum, low heat of fusion, dielectric constant, etc.

CMCs for sodium alkylsulfates and alkyltrimethylammonium bromide (C_nTAB) in hydrazine and water are summarized in table 2. These CMCs in hydrazine were determined by sodium and bromide NMR measurements and changes in fluorescence spectrum and are about two times larger than those in water. Krafft temperatures, the lowest temperatures at which a CMC

Table 1. *Comparison of hydrazine and water*[a]

Property	N_2H_4	H_2O
mp, °C	1.69	0.0
bp, °C	113.5	100.0
T_c, °C	380	374.2
P_c, atm	145	218.3
density at 25 °C, g cm^{-3}	1.0036	0.9971
η at 25 °C, P	0.00905	0.008904
γ at 25 °C, dyn cm^{-1}	66.7	72.0
ΔH_f at mp, cal mol^{-1}	3025	1440
ΔS_f, cal mol^{-1} K^{-1}	11	5.3
ΔH_v at bp, cal mol^{-1}	9760	9720
ΔH_v at 25 °C	10700	10500
ΔS_v at bp, gibbs mol^{-1}	25.2	26.1
M (gas), D	1.83–1.90	1.85
ϵ at 25 °C	51.7	78.3
η_D	1.4644	1.3325
sp. conductance at 25 °C, mho cm^{-1}	3×10^{-6}	5×10^{-8}
ion product, mol^2 cm^{-6}	2×10^{-25}	10^{-14}
C_p (liquid), cal mol^{-1} K^{-1}	23.62 (298 K)	17.98 (298 K)
C_p (solid), cal mol^{-1} K^{-1}	15.3 (274.7 K)	8.9 (273 K)

[a] Compiled by H. S. Frank and R. Lumry. Data for hydrazine taken from ref. 2.

Table 2. *Critical micelle concentration in hydrazine and water*

Surfactant	T(°C)	10 H_4N_2 (M)	10 H_2O (M)
C_8OSO_3Na	35	3.0	1.3
$C_{10}OSO_3Na$	25	0.58	0.33
	35	0.80	0.34
	45	1.14	0.44
$C_{12}OSO_3Na$	35	0.22	0.0857
	45	0.30	0.0910
$C_{10}TAB$	25	1.5	0.65
	35	2.2	
	45	2.5	
$C_{12}TAB$	35	0.64	
	45	0.83	

Table taken from ref. 2.

can be detected, are about 10–15 °C higher than in water. Figure 2 compares the change of CMC with surfactant chain length in water and hydrazine. The slopes of these lines permit the free energy of transferring a methylene group from the solvent to the micelle to be determined and yield −720 and −705 cal mol^{-1}, respectively. Thus these two solvents display almost identical solvophobic properties. Figure 3 compares the change of CMC for

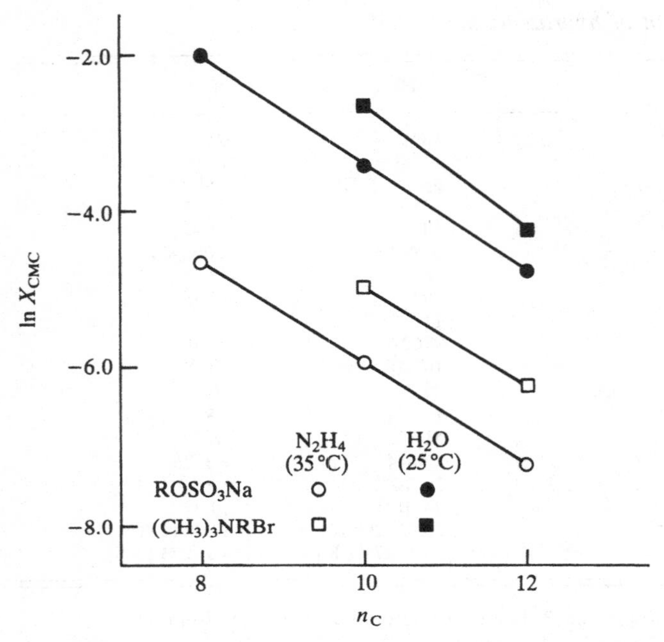

Figure 2. Ln X_{CMC} vs carbon number, n_C, for sodium alkyl sulfates and C_nTABs in hydrazine at 35 °C and water at 25 °C. The similar slopes in hydrazine and water indicate that the two solvents display almost identical solvophobic properties. From ref. 2.

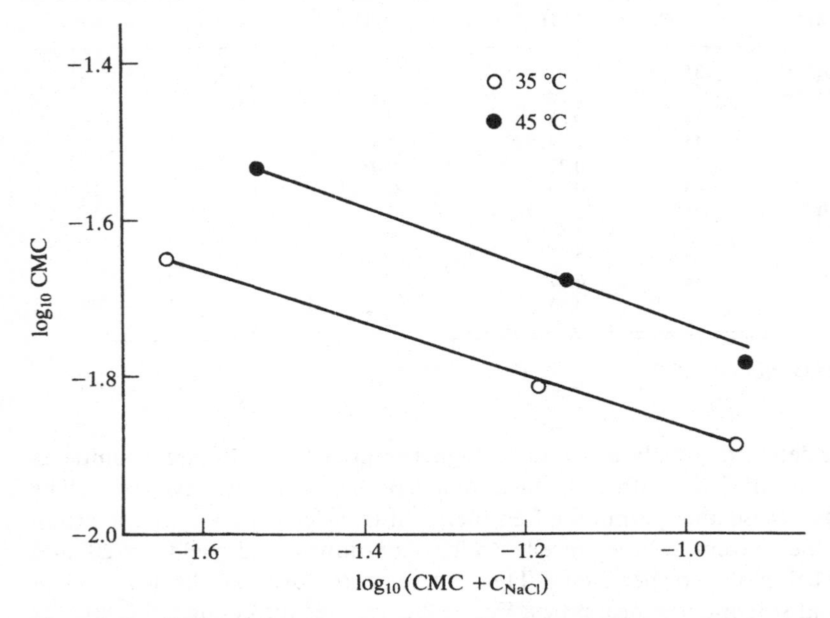

Figure 3. Ln CMC vs. ln (CMC + C_{NaCl}) for sodium dodecyl sulfate in hydrazine at 35 and 45 °C. Such plots allow the degree of counterion binding to be estimated, and give a value of 0.8 which is comparable to that in water. From ref. 2.

Table 3. *Comparison of thermodynamics of micelle formation and hydrocarbon transfer in water and hydrazine*

	Surfactant	T (°C)	$-\Delta G$(HP) (kcal mol^{-1})	$-\Delta H$(HP) (kcal mol^{-1})	ΔS(HP) (cal mol^{-1} K^{-1})
H_2O	$C_{14}TAB$	25	10.5	3.0	25
		95	11.5	12.4	−2.5
		166	10.6	15.6	−12
H_4N_2	$C_{12}TAB$	35	7.1	11.0	−13
	$C_{12}OSO_3Na$	35	8.6	14.0	−18

sodium dodecylsulfate with added sodium chloride in water and hydrazine. The slope of these lines permits us to estimate the degree of counterion binding and gives a value of 0.8 which is comparable to that in water.

Analysis of the temperature dependence of the CMC with equation (1) provides the free energy, entropy and enthalpy values given in table 3. In hydrazine, the negative entropy reflects the order accompanying the transfer and orientation of the hydrocarbon chains in the micelle. We can conclude that micellization is driven entirely by the enthalpic changes associated with this microphase separation; the water results will be discussed below. Note that this analysis assumes that the micellar aggregation numbers in hydrazine are identical to those in water.

With aggregation numbers between 40 and 70, the dressed micelle model [11] is relatively insensitive to changes in this quantity. We note in passing that CMCs for sodium dodecylsulfate have been determined in hydrazine–water mixtures.[3] Unlike most water co-solvent systems,[12] the changes in free energy, enthalpy and entropy are completely linear across the entire solvent composition range.

We next consider the low melting (mp = 16 °C) fused salt EAN.[4] With three protons on the cation and three basic oxygens on the anion, this solvent is capable of forming a three-dimensional structure. Table 4 compares the thermodynamics of transferring gases from cyclohexane to water and EAN.[13] The strikingly negative values of ΔH° and ΔS° that have dominated much of our thinking about hydrophobic effects in water are also seen in EAN, but the magnitudes in EAN are reduced by one-half. This clearly demonstrates that such anomalous entropic and enthalpic changes are not unique to aqueous solutions. As we will develop in more detail below, the basis for differentiating between water and other solvents rests not in the first derivatives of the free energy, but rather in the secondary derivatives like heat capacity, expansivity, etc. Mirejovsky and Arnett [14] discuss this point with regard to EAN.

CMCs have been determined in EAN [4] by surface tension, light scattering and neutron scattering measurements for the nonionic surfactant,

Table 4 *Thermodynamics of transfer of nonpolar gases from cyclohexane to EAN and water*

		$\Delta G°$ (kcal)	$\Delta H°$ (kcal)	$\Delta S°$ (cal K^{-1})
Kr	FS	1.6	−0.9	−9
	H_2O	2.8	−2.9	−19
CH_4	FS	1.6	−0.5	−4
	H_2O	2.9	−2.7	−18
C_2H_6	FS	2.0	−1.0	−10
	H_2O	3.9	−2.1	−20
C_4H_{10}[a]	FS	(3.61)	(−5.71)	(−31.3)
	H_2O	(6.35)	(−6.21)	(−42.1)

[a] The values for butane refer to the transfer from the gas phase to the fused salt and to water.

Table 5. *CMCs in EAN at 50 °C*

Surfactant	CMC (mole fraction)
$(C_{12}H_{25})N(CH_3)_3Br$	2.30×10^{-2}
$(C_{14}H_{29})N(CH_3)_3Br$	6.97×10^{-3}
$(C_{16}H_{33})N(CH_3)_3Br$	2.23×10^{-3}
	2.0×10^{-3}[a]
$(C_{14}H_{29})C_5NH_5Br$	6.49×10^{-3}
	8.6×10^{-3}[b]
$(C_{16}H_{33})C_5NH_5Br$	1.69×10^{-3}
$(C_{18}H_{37})C_5NH_5Br$	5.69×10^{-4}
Triton X-100[c]	6.11×10^{-3}
	5.93×10^{-3}[d]

[a] Determined by viscosity measurements.
[b] Estimated from light scattering measurements, 6.
[c] Using a molecular weight of 624 g mol^{-1}, for Triton X-100.
[d] Determined at 21 °C.

Triton X-100, and the cationic alkyltrimethylammonium and alkylpyridinium bromides (table 5). In general, the CMCs are 10–15 times larger than in water and Kıafft temperatures are higher by about 15 °C. From the change in CMC with alkyl chain length, we obtain the free energy of transferring a methylene group as −405 cal mol.$^{-1}$ Since EAN is the equivalent of an 11 M ionic solution, analysis with the dressed micelle model is clearly inappropriate. The phase separation model, however, is applicable, and analyis of the temperature dependence of the CMC for Triton X-100 gives $\Delta G = -3.0$ kcal mol^{-1}, $\Delta H \sim 0$ and $\Delta S = 10$ cal mol^{-1} deg^{-1} for hydrocarbon chain transfer. The corresponding values in water are $\Delta G = -7.3$ kcal mol^{-1}, $\Delta H = 2.1$ kcal mol^{-1} and $\Delta S = 32$ cal mol^{-1} deg^{-1}.

The aggregation numbers in EAN for tetradecylpyridinium and hexadecylpyridinium bromide are 17 and 26, about one-third of the value observed in water.[5] Simple geometric packing for a sphere gives head group areas of 103 and 97 Å^2, respectively. There are two possible explanations for these small micelles in EAN. The first is that since the free energy costs associated with exposure of hydrocarbon to solvent are less in EAN, the micelles can be smaller. The second is that the ethyl groups of ethylammonium ions are also incorporated into the micelles as a co-solvent. Since all species in the solution are ionic, incorporation of a positive charge cosurfactant would have little effect on the intramolecular interactions.

Liquid crystals formed from phospholipids have been characterized in EAN.[15] In general, the spacing between phosphatidylcholine headgroups is larger than that observed in water, compatible with the diminished cost associated with hydrocarbon-solvent exposure.[16] The report of micelle formation by dipalmitoylphosphatidyl-choline based on interpretations of Raman scattering data [17] is not compatible with the bilayer spacing obtained from x-ray scattering data.[16]

Partial molar volumes and heat capacities have been reported for EAN–water mixtures.[18] In such mixtures, the enzyme, alkaline phosphatase, retains considerable activity up to 60% EAN (v/v) and is stable to brief exposures to solutions as high as 80% EAN. [19]

Much less information is available on micelles in formamide and ethylene glycol.[6] The Krafft temperatures and CMCs in both solvents are generally considerably higher than in water.[20] Scattering data to establish the size and properties of the micelles are needed. Liquid crystal formation and microemulsions employing formamide [21] and ethylene glycol [22] have been reported and characterized.

These studies of micellization in nonaqueous solvents establish that the origin of aggregation phenomena is mainly the solvophobicity of the hydrocarbon chains in hydrogen bonding solvents.

2.3 *Water*

Every scientist who makes measurements on aqueous solutions at elevated temperatures implicitly understands a great deal about the unusual solvophobic properties of water. In order to prevent bubble formation, aqueous solutions must be degassed; i.e. at room temperature the solubility of nonpolar gases initially decreases with increasing temperature. This is shown quantitatively in figure 4 where the solubilities of alkylbenzenes are plotted vs the reciprocal temperature.[23] At elevated temperatures, where the unique structural properties of water disappear and it behaves like a normal hydrogen bonded solvent, the solubility increases with increasing temperature in a manner typical for most solvents. However, at lower temperatures, the behavior in water is more complex, the solubility goes

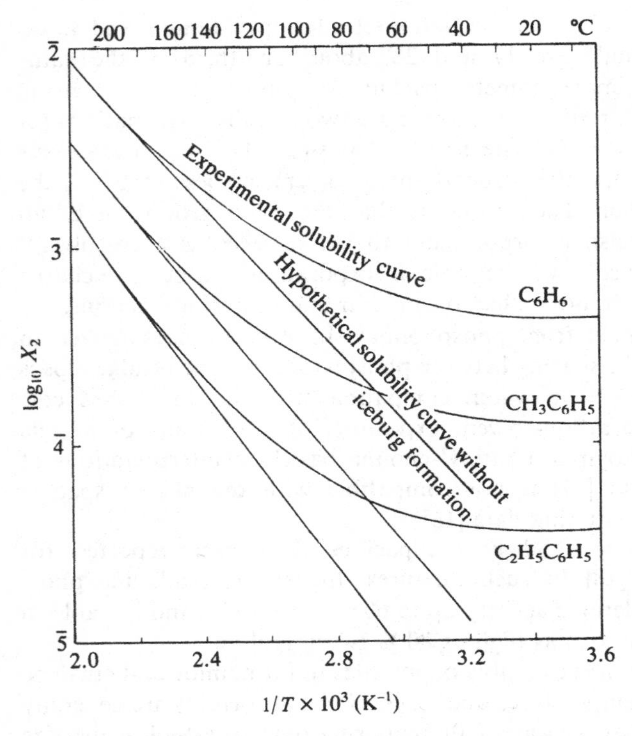

Figure 4. Solubility of alkyl benzenes in water (expressed in mole fraction units) as a function of reciprocal temperature. Hypothetical curves (straight lines) give solubilities of hydrophobic molecules in 'structureless' water. Experimental curves demonstrate that water structure at lower temperatures *increases* the solubilities of hydrophobic moieties. From ref. 23.

through a minimum and at lower temperatures actually increases with decreasing temperature. A linear extrapolation of the high temperature behavior to 25 °C provides an estimate of the solubility in water in the absence of the special structural properties of water which become pronounced with decreasing temperature. Comparisons of the extrapolated and measured solubilities clearly show the extent to which solubilities are augmented in low temperature water. Shinoda pointed out the implications of this in 1978.[23] The conclusion is clear and unambiguous: judged by the criteria of how the free energy, i.e. $RT \ln X_2$ (where X_2 is the mole fraction solubility) varies with temperature, hydrocarbon–water structural interactions below 90 °C actually lead to an *increase* in the solubility of nonpolar molecules in aqueous solutions over that in other polar solvents. This explains the origin of the deviation of water from the linear free energy vs $\gamma/V^{\frac{1}{3}}$ behavior of other polar solvents (see figure 1).

This conclusion directly contradicts the traditional inferences drawn from

Table 6. *Thermodynamic function for transfer of 1 mol of argon from cyclohexane to water and to hydrazine*

Step	ΔG, (*kcal* mol^{-1})	ΔH, (*kcal* mol^{-1})	ΔS, (gibbs mol^{-1})
(1) Ar(X_2 = 1 in C_6H_{12}) = Ar(X_2 = 1 in H_4N_2)	2.84	2.27	−1.9
(2) Ar(X_2 = 1 in H_4N_2) = Ar(X_2 = 1 in H_2O)	−0.36	−4.95	−15.4
(3) Ar(X_2 = 1 in C_6H_{12}) = Ar(X_2 = 1 in H_2O)	2.48	−2.68	−17.3

entropic and enthalpic changes that are based on the analysis of Frank and Evans.[24] They analyzed the thermodynamics of transferring rare gases from the vapor phase to water and other solvents. For our purposes it is more convenient to use liquid cyclohexane as a reference rather than the gaseous state and we will focus on argon as a solute (table 6, line (3)). The positive free energy simply reflects the truism that oil and water don't mix. The negative entropies and enthalpies were surprising and different from all other data then available.

Frank and Evans concluded that nonpolar groups induced an increase in the water structure in their immediate vicinity. This occurs by formation of more or stronger hydrogen bonds; the negative enthalpy reflects the exothermic nature of the process and negative entropy arises from the ordering associated with hydrogen bond formation. The existence of solid stoichiometric clathrates,[25] in which each nonpolar group is completely caged inside a water cavity, provides well-defined crystalline structures. In the liquid state, remnants of these water structural cages persist even upon melting. Thermal fluctuations result in flickering clusters that continuously form and disperse around nonpolar groups.

As more complex situations encountered in biological and industrial processes were considered,[26] Frank and Evans' observations on the role of 'hydrophobic interactions' became the guiding light. However, conclusions regarding the effect of water structure on the solubility of nonpolar groups as judged from the change in solubility with temperature or from the ΔH and ΔS values apparently contradict one another. We can understand this apparent paradox by considering 'hydrophobic processes' in water over an extended temperature.

We consider first the micellization of C_{14} TAB in water from 25 to 160 °C.[27] Over this rather large temperature range, in which water becomes a 'normal' polar hydrogen bonding solvent, the CMC increases by a factor of 10 while the aggregation number decreases from 72 to 8.[28] Analysis of the data with the dressed micelle model gives the values of $\Delta G°$, $\Delta H°$ and $\Delta S°$ for hydrocarbon transfer chains in figure 5. Thus the 'hydrophobic' driving force is almost independent of temperature.

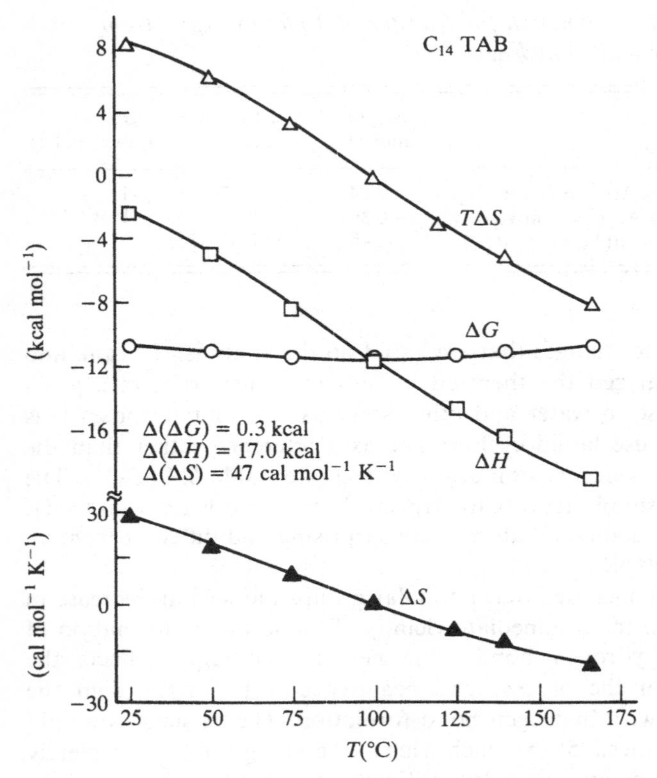

Figure 5. Thermodynamics of the micellization of C_{14}TAB obtained using the dressed micelle model. Despite large changes in ΔH and $T\Delta S$, the free energy of micellization (ΔG) remains almost constant across the entire temperature range from 25 to 160 °C.

The very large enthalpy and entropy changes compensate so as to leave the free energy almost invariant. At high temperatures, the large negative entropy reflects the ordering associated with transferring the hydrocarbon chains out of water and into the confines of a micelle and simultaneously orienting the headgroup at the water interface. In other words, at high temperatures micellization is driven entirely by enthalpic changes. These same changes must also occur at 25 °C, but ΔH and ΔS are decorated by water structural changes that have only a small influence on water–nonpolar interactions.

The results in water at high temperatures parallel those in hydrazine at 35 °C (table 3) and emphasize the fact that the solubility of nonpolar groups in aqueous solution is mainly a matter of solvophobicity. In order to understand the thermodynamics below 90 °C, one must consider the nature of compensation in aqueous solutions.

2.4 *Compensation*

Over the years the subject of compensation has provoked considerable puzzlement and exasperation. This is due in part to the complexity of the subject, in part to a lack of a clear understanding of its basis and in part to the analysis of data which always seem to yield excellent fits to compensation plots independent of the quality of the original data.[29] Recently a firm basis for discussing compensation has emerged as a consequence of the analysis by Benzinger.[30,31] In addition, the precautions necessary in interpreting data have been realized.

A useful starting point for understanding the thermodynamics of compensation is the Gibbs expression for the free energy

$$\Delta G = \Delta H - T\Delta S \tag{2}$$

The enthalpy at a given temperature T can be written as

$$\Delta H = \Delta H^{\circ} + \int_0^T \Delta C_p \mathrm{d}T \tag{3}$$

where ΔH° is the enthalpic change at absolute zero and the integral gives the heat necessary to raise products and reactants to the temperature T.

In evaluating the entropic term,[32] it is useful to integrate by parts

$$\int \Delta S \mathrm{d}T = T\Delta S - \int T(\partial \Delta S/\partial T)\mathrm{d}T \tag{4}$$

where $\partial \Delta S/\partial T \equiv \Delta C_p/T$. Combination and rearrangement yields

$$T\Delta S = \int \Delta S \mathrm{d}T + \int \Delta C_p \mathrm{d}T \tag{5}$$

Combining equations (2), (3) and (4) gives

$$\Delta G = [\Delta H^{\circ} + \int \Delta C_p \mathrm{d}T] - [\int \Delta S \mathrm{d}T + \int \Delta C_p \mathrm{d}T] \tag{6a}$$

$$\Delta G = \Delta H^{\circ} - \int \Delta S \mathrm{d}T = \Delta H_m - T\Delta S_m \tag{6b}$$

which shows that there are components of ΔH and $T\Delta S$ which always cancel. These compensated parts are the change in enthalpy associated with the heat needed to increase the temperature of the system and the corresponding entropy change associated with it. At constant temperatures these are not available to do work in the surroundings and thus cancel out in evaluating the free energy, yielding

$$\Delta G = \Delta H_m - T\Delta S_m \tag{7}$$

Notice that the remaining terms are the enthalpic change at absolute zero and the entropy change which can be related to $k\ln \Omega$ and are labeled with a subscript m for the motive or driving contribution to the free energy.

This is second law compensation as developed by Benzinger. It demonstrates that there is no simple relationship between the free energy and the sign or magnitude of the corresponding enthalpy or entropy. Thus, one can never draw inferences about *why* a process occurs from a knowledge of ΔH or ΔS. These conclusions are evident in Slater's quantum mechanical calculations for crystalline metals at low temperatures, although he made no explicit comments on the subject.[33] It is illustrated by Frank [34] who calculated the separate contributions to the entropy. For example, at 298 K the compensated and motive entropies in cal mol^{-1} deg^{-1} are 5.4 and 9.8 for lead, 4.8 and 6.8 for gold and 4.0 and 4.1 for copper.[29,34,35] Thus, the compensated contributions range from 36 to 50%. For more complex systems where changes in crystalline structure or phase are encountered, numerical evaluation becomes very difficult because it is impossible to split the enthalpic and entropic contributions into motive and compensated parts.[29]

The discussion outlined above focuses on second law compensation. There are, however, at least three additional sources of compensated behavior. These include processes occurring in water, in solution processing of many types and in the broad area of linear free energy relationship. These have been analyzed by Lumry and Gregory[29] and the reader is referred to their article for a detailed discussion.

We can illustrate many of the features of 'hydrophobic' compensation in aqueous solution by comparing the solubility of argon in water and in hydrazine (table 6). The idea is to divide the dissolution process in water into two imaginary steps. In the first step, we create a hole in water and insert the argon with the proviso that water is inhibited and cannot structurally respond to the presence of the nonpolar group. Because of the physicochemical similarities between hydrazine and water (particularly the enthalpy of vaporization and the surface tension), hydrazine is proposed as a model for inhibited water (table 6, line (1)). In the second step, we relax the inhibitions, which corresponds to the transfer of argon from hydrazine to water (table 6, line (2)). Large negative changes in ΔH and ΔS occur, accompanying the formation of flickering clusters, but they are compensated so that the change in free energy is very small. While this is a heuristic argument, it illustrates the point in a useful way.

Relative to other liquids, water possesses an enormous diversity of fluctuating states that display comparable free energy, but vastly different entropy and enthalpy. A nonpolar molecule imposes on water a succession of these states in a way that actually increases the molecule's solubility at low temperature. In such a situation, the compensated contributions to the enthalpy and entropy dominate and the measured values provide no useful insight as to why processes happen. Whenever a large number of molecular configurations of comparable free energy exists, we find enthalpy–entropy compensation.[29]

3. Structure of amphiphilic assemblies

3.1. *Thermodynamic theories of aggregation*

A priori prediction of aggregate size and shape becomes possible if we are provided with a theoretical framework through which molecular parameters (e.g. hydrophobic volume and chain length, headgroup area) and field variables (e.g. temperature, ionic strength) are related to microstructure. Through free energy minimization arguments, such theories attempt to determine the 'optimal aggregate(s)' for a given set of molecular parameters and field variables.[36–40] In dilute solution, the free energy of surfactant self-assembly is assumed to be made up of three terms:[40] (1) a favorable, hydrophobic term ($\Delta g(\mathrm{HP}) < 0$), due to sequestering of the hydrocarbon chains into the interior of the aggregates, (2) a surface term, which reflects the opposing tendencies of the surfactant headgroups to crowd close together (to minimize hydrocarbon–water contacts) and spread apart (due to electrostatic repulsion, hydration and steric hindrance), and (3) a packing term, which, at its simplest level, requires that water be excluded from the hydrophobic interior of the aggregate and therefore limits the geometrically inaccessible forms available to the aggregate. For example, since water is excluded from the hydrophobic interior, spherical and cylindrical micelles cannot have radii larger than the length of the hydrocarbon chain of the amphiphile. The surface and packing terms will take on different functional forms for each postulated aggregate geometry, and the optimal aggregate is determined by the form giving the minimum free energy for a given set of conditions.

For dilute solutions in which interactions between aggregates are not important, these ideas are conveniently subsumed into the surfactant parameter v/al,[36,37] where v is the volume of the hydrophobic portion of the surfactant molecule, l is the length of the hydrocarbon chains and a is the effective area per headgroup. For systems in which $v/al < \frac{1}{3}$, theory demonstrates that the most likely aggregate is the small spherical micelle. Cylindrical micelles are predicted for $\frac{1}{3} < v/al < \frac{1}{2}$, bilayers, vesicles and other large aggregates for $\frac{1}{2} < v/al < 1$ and inverted structures for $v/al > 1$. Thus, the surfactant parameter is a measure of local curvature; small values of v/al correspond to highly curved structures while values near unity correspond to large, nearly planar structures.

The approximate nature of this self-assembly theory is obvious; in spite of this, it provides valuable insight regarding the effect of changes in solution conditions and molecular structure on aggregate size and shape. For example, it is well known that the addition of salt to cetyltrimethylammonium bromide (CTAB) micelles induces a change from spherical to cylindrical micelles.[41] At low ionic strength the surfactant parameter of CTAB is in the spherical micelle range, and the experimentally accepted shape for CTAB micelles is indeed a sphere. Upon addition of salt, we expect the area per headgroup a to decrease (due to electrostatic screening of the repulsive forces) and v and l to remain constant. As a result, the surfactant parameter increases into the cylindrical range. Other changes in solution (or molecular) conditions that affect headgroup repulsions will also have predictable structural effects. These are all based on changes in the effective headgroup area (and consequently aggregate curvature), as we will see below.

The effect of adding a second hydrophobic chain to the polar headgroup is also predicted by our simple theory. Amphiphiles with a single hydrophobic chain have surfactant parameters less than $\frac{1}{2}$ and are constrained, at least in dilute solution, to form micellar aggregates. The second hydrocarbon chain doubles v while a and l remain essentially the same, thus in effect doubling v/al into the range $\frac{1}{2}$–1. Hence, double-chain surfactants tend to form structures of inherently lower curvature (e.g. vesicles and liposomes) than their single-chain counterparts. Unfortunately, attempts to predict actual aggregate shape in the range $\frac{1}{2} < v/al < 1$ are frustrated by the fact that curvature energies are small, and thus a wide range of structures, differing in energy by at most kT, coexist. These structures include, as we shall see, liquid crystalline liposomes, unilamellar vesicles, coiled microtubules and multilayered tubules, to name just a few.

We are now in a position to interpret experimental data that attempt to correlate changes in aggregate size and shape with changes in molecular and field parameters in terms of local curvature. We start with data on single-chained surfactants since these compounds form well-characterized micellar aggregates to which detailed, quantitative theories of aggregation can be applied. The conclusions drawn from this discussion will be of use in our interpretation of the more complex behavior of double-chained surfactants.

Recent studies by Roelants and DeSchryver,[42] Beesley and Evans,[28] and Miller, Magid and Evans,[43] nicely summarize some of the parameters affecting ionic micelles.

(1) *Chain length.* Simple geometric considerations apply to spherical micelles. Thus,

$$Nv = \tfrac{4}{3}\pi R^3 \tag{8}$$

and

$$Na = 4\pi R^2 \tag{9}$$

where N is the aggregation number and R is the radius of the micelle. Obviously, an increase in chainlength will increase the radius of the micelle, resulting in a larger aggregation number. Table 7 lists the aggregation numbers of alkyltrimethylammonium chloride surfactants with 12 (C_{12} TAC), 14 (C_{14}TAC) and 16 (C_{16}TAC) carbon chains.[42]

(2) *Ionic strength.* As already mentioned, addition of salt to sodium dodecylsulfate (SDS) causes a decrease in headgroup repulsion, resulting in decreased curvature and increased aggregate size (see also equation (9)). This

Table 7. *Aggregation numbers for alkyltrimethylammonium chlorides as a function of chainlength*

Number of carbons	*N* (aggregation number)
12(C_{12}TAC)	47
14(C_{14}TAC)	66
16(C_{16}TAC)	81

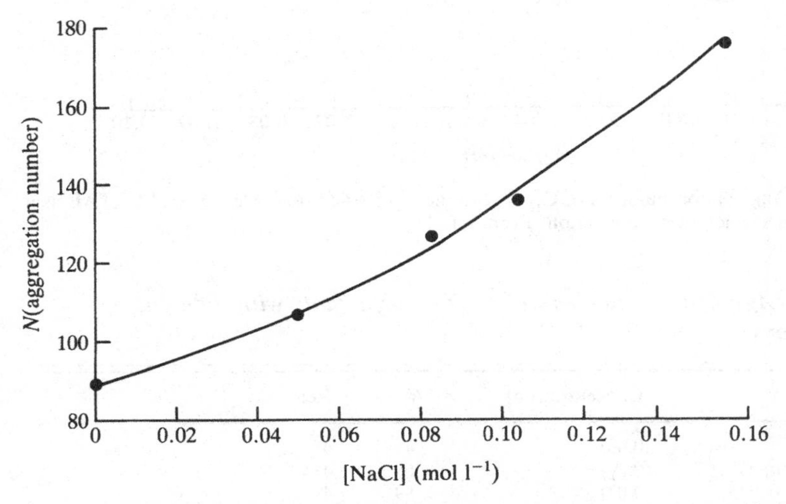

Figure 6. Aggregation numbers of C_{16}TAC micelles as a function of sodium chloride concentration. The growth of micelles with added salt is seen in most ionic surfactants. Data from ref. 42.

behavior is typical of all ionic surfactants, as is illustrated by addition of NaCl to C_{16}TAC micelles [42] (see figure 6).

(3) *Surfactant concentration*. The size of ionic micelles generally increases with concentration. Data for hexadecyltrimethylammonium (C_{16}TA) surfactants and C_{14}TAB micelles are given in figure 7.[42,43] This increase in micelle size is due to the increase in solution ionic strength as well as the added influence of intermicellar interactions, which become important at increased surfactant concentration.

(4) *Counterion*. Dramatic changes in surfactant aggregation behavior can be induced by changing the counterion of ionic surfactants. For example the aggregation numbers of hexadecyltrimethylammonium surfactants are given in table 8.

The over two-fold decrease in aggregation number that occurs when the bromide counterion is replaced by the hydroxide ion parallels a large increase

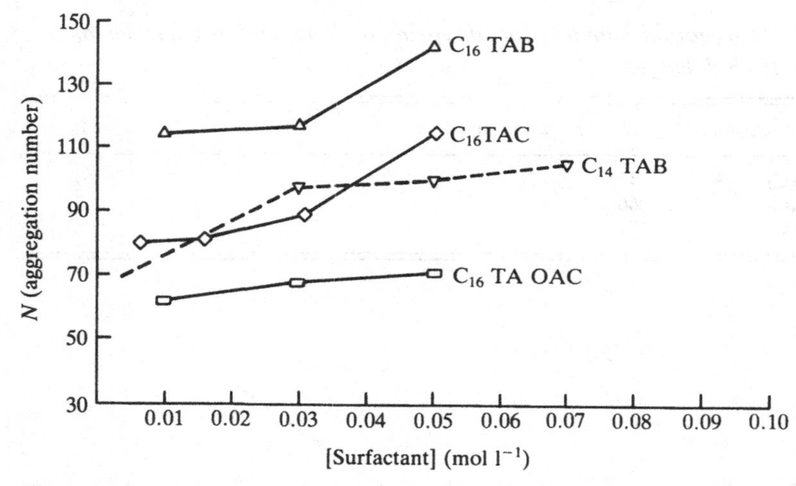

Figure 7. Aggregation numbers of C_{16}TA bromides, chlorides and acetates and C_{14}TAB as a function of surfactant concentration. From ref. 43.

Table 8. *Aggregation numbers of $C_{16}TA$ surfactants with different counterions*

Counterion	(Concentration)	N	Ref.
OH^-	(0.05)	46	44
OAc^- (acetate)	(0.03)	68	43
Cl^-	(0.03)	89	42
Br^{-1}	(0.03)	117	43

in the degree of ionization of the micelle. Thus, the experimentally determined ionization degree (α) of C_{16}TAB is about 0.22 [45] (meaning 78% of the counterions are 'bound' to the surface of the micelle) while that of the corresponding hydroxide surfactant (C_{16}TAOH) ranges from twice (0.49) [45] to three times this value (0.70),[46] depending on the technique used to measure α.

It seems obvious, then, that the origin of the decrease in micelle size (and hence increase in local curvature) when bromide is replaced by hydroxide is an increase in effective headgroup area; because the bromide ions are more tightly 'bound,' they are more effective at screening headgroup repulsions than the hydroxide ion. Chloride and acetate ions have intermediate abilities to screen headgroup repulsions.

The dressed micelle model [10] can account for these counterion effects only if the ions that 'bind' more tightly to the micelles are, on averge, *closer* to the micelle than those that 'bind' less tightly. This is reasonable since counterions like hydroxide and acetate are more strongly associated with water (i.e. have larger hydrated radii) and will sit out further from the micelle surface than the less strongly hydrated chloride and bromide ions. One calculation shows that the experimental data are consistent with the hydroxide ion sitting about 4 Å further from the micelle surface than the bromide ion.[47]

With anionic surfactants, the nature of counterion binding can be changed by complexing alkali metal ions with macrocyclic polyethers or cryptates.[48] For example, combining of sodium dodecylsulphate (SDS) with the C222 cryptate (4,7,13,16,21,24-hexaoxa-1,10-diazabicyclo-(8,8,8)hexacosane) results in the micellar aggregation number decreasing from 60 to 40. The effect of added C222 on SDS is most dramatic with added salt. The usual increase in aggregation number with added salt is completely suppressed, for example, the aggregation number of 40 is also found in 0.3 M Na(C222)Cl. Surface tension measurements establish that the area per dodecylsulfate ion at the air–water interface increases from 65 Å^2 to 140 Å^2 upon counterion complexation. The formation of the large sodium cryptate inclusion complex (diameter $\approx$ 10 Å) draws counterions away from the sulfate headgroup.

Counterion complexation by macrocyclic compounds shows great ionic specificity. The complexing constants ($\log_{10}K_{st}$) for C222, complexing Me_4N^+, Li^+, Na^+ and K^+ ions are 1, < 2, 4.7 and 5.2 respectively, while that for C221 (4,7,13,16,21-pentaoxa-(1,10-diazabicyclo 18,8,5)tricosane) complexing for Li^+ is 2.1. The effect of the cryptates on micellar properties parallels the degree of counterion complexation. With complexation, Krafft temperatures are lowered and potassium, calcium and barium dodecylsulfate micellar solutions can be characterized at 25 °C.

The insight gained from counterion studies on single-chained surfactants is that more highly hydrated, effectively larger, counterions result in decreased headgroup screening, increased local curvature and smaller aggregates. This insight will be useful in our discussion of double-chained surfactants.

(5) *Temperature.* Because of added experimental complexity, the effect of temperature on the aggregation behavior of ionic, single-chain surfactants has not been studied extensively, especially at high temperatures (> 90 °C). The results of one of the few such experiments, a light scattering study of C_{14}TAB up to 160 °C,[28] were reported in the first section; micellar size at the CMC decreases from 70 at 25 °C to less than 10 at 160 °C. By applying the dressed micelle model [10] to these data, we find that the hydrophobic free energy (Δg(HP)) remained roughly constant over the entire temperature range (figure 5). This is entirely consistent with gas solubility data which demonstrate that the solubility of inert gases at low temperatures is equal to that at high temperatures ($\sim$ 160 °C), with a minimum at $\sim$ 90 °C.[23] Our intuition, which tells us that micelles get smaller with increasing temperature because of diminished driving force for aggregation, is clearly incorrect. Of

course, as the temperature rises even higher, entropic effects will eventually dominate and the surfactant will take the form of randomly dispersed, unaggregated monomers. We can try to invoke the by now familiar argument of increased headgroup repulsion to explain the increase in local curvature with temperature, but we require a plausible mechanism to explain such a phenomenon. Hydration arguments, which work well to explain counterion effects, require an *increase* in hydrated radius with temperature to be consistent with the data. This seems unreasonable.

The electrostatic model for headgroup repulsions implies that we can use double-layer theory to describe their strength. Thus, the important scaling quantity is the Debye length (κ^{-1}); large values of κ^{-1} correspond to large repulsions. At constant ionic strength, κ^{-1} is proportional to $(\epsilon_I kT)^{\frac{1}{2}}$ where ϵ_I is the dielectric constant. The subscript I reminds us that the headgroups lie in the interface between bulk water and the oil-like interior of the micelle, and that we should use the effective dielectric constant of this interfacial region to describe interactions between headgroups. A recent spectroscopic determination of the effective dielectric constant of micelle–water interfaces between 15 and 85 °C [49] reveals that, unlike bulk dielectric constants which decrease with temperature, ϵ_I remains constant over this entire temperature range. The quantity $\epsilon_I kT$ therefore *increases* with temperature, with the result that headgroup repulsions (and local curvature) also increase with temperature.

In the derivation of Debye–Hückel, Gouy–Chapman or dressed micelle theory, all of the temperature dependencies arise from the Boltzmann equation describing the distribution of ions around a charge surface. Thus

$$C_i = C_{i0} \exp(-\psi z_i e/kT) \tag{10}$$

where C_i is the concentration of ions at a distance from the surface in which the electrostatic potential is ψ, C_{i0} is the bulk concentration of ions, z_i is the valency (+ or −) of the ions and e is a unit charge. Calculations using the dressed micelle model and aggregation data from $C_{14}TAB$, indicate that the surface potential of the micelles is only a weak function of temperature;[43] in other words, thermal effects (kT) dominate equation (10). Thus, increases in thermal motion (kT) tend to smear out the accumulation of counterions about a charged surface and make the structure of the double layer more like the random distribution of ions in the bulk. The result is a decrease in electrostatic screening from the counterions and an increase in electrostatic repulsion. An appeal to the dressed micelle model reveals that the concept of counterion binding arises naturally from this Boltzmann accumulation of counterions about oppositely charged surfaces. Hence, the result of increased temperature is an increase in the thermal motion of the counterions (in the interfacial region of the micelle) resulting in decreased 'ion binding,' increased headgroup repulsions and decreased aggregate size.

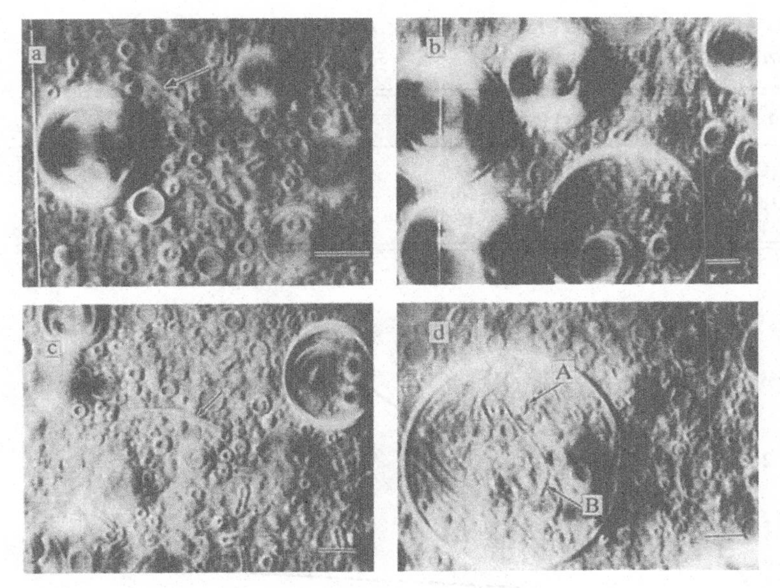

Figure 8. Video micrographs of 0.01 M $2C_{12}N2C_1Br$. Bar = 5 μm.

We will explore temperature effects on aggregate structure further when we deal with double-chained surfactants.

As stated above, most double-chained amphiphiles have surfactant parameters that lie in the range $\frac{1}{2} < v/al < 1$. Aggregates of such amphiphiles, therefore, tend to be of low local curvature, resulting in large structures with many different shapes. Video enhanced microscopy (VEM) and cryo-transmission electron microscopy (cryo-TEM) reveal [50] that such diverse structures as liquid crystalline liposomes, microtubules, coiled microtubules inside vesicles, vesicles encapsulating smaller vesicles and even, in some cases micelles, are formed by these compounds. Micelles are too small to be imaged by direct techniques such as VEM and cryo-TEM but are detected in some such dispersions by a modification of the fluorescence quenching technique.[51] A sampling of some of these aggregates, formed by the cationic didodecyldimethylammonium bromide ($2C_{12}N2C_1Br$), is given in figure 8. Similar structures are formed by the anionic sodium 8-hexadecyl-ρ-benzenesulfonate (SHBS). See ref. 50 for more VEM and cryo-TEM micrographs of such aggregates.

These structures are also formed by phospholipid amphiphiles in living systems. In this case, however, interconversion of structural types is accomplished by controlled responses to physiological needs, not random thermal motion. For example, small exocytotic vesicles are created from the bilayer of the cell membrane when bioactive material and metabolic products

Table 9. *Aggregation numbers for dialkyldimethylammonium acetate surfactants (0.01 m)*

Surfactant	[NaOAc] (M)	*N*
$2C_{14}N2C_1OAc$	—	84.9
$2C_{12}N2C_1OAc$	—	56.7
$2C_{12}N2C_1OAc$	0.1	94.6

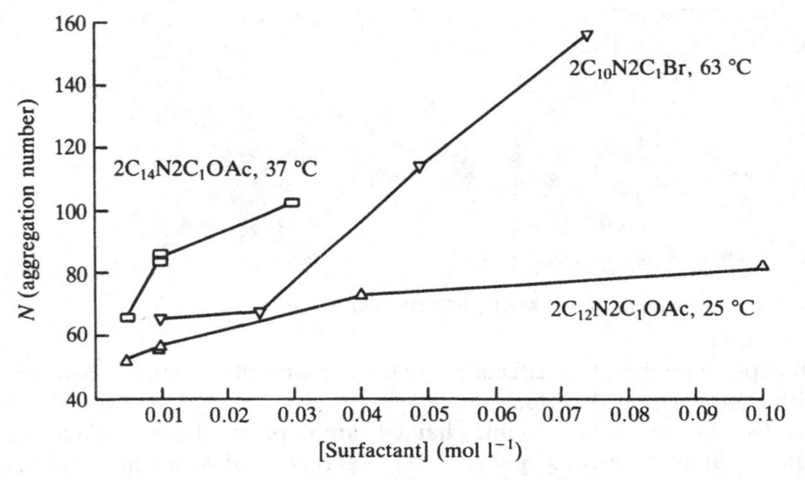

Figure 9. Aggregation numbers of dialkyldimethylammonium acetates and bromides ($2C_{12}N2C_1OAc$, 25 °C, $2C_{14}N2C_1OAc$, 37 °C and $2C_{10}N2C_1Br$, 63 °C) as a function of surfactant concentration. From ref. 51.

are carried away from the cell. In order to begin to understand such a system, we start with primitive, single component, ionic double-chained surfactants.

The effect of counterion substitutions on single-chained surfactants is to change the size of the micelle through hydrated ion-mediated headgroup interactions. With double-chained surfactants, such substitutions can cause much more dramatic structural changes. For example, $2C_{12}N2C_1Br$ and $2C_{14}N2C_1Br$ form turbid dispersions of liposomes and other complicated structures at concentrations as low as 0.05 wt%.[44,50] By replacing the bromide ion with an acetate or hydroxide ion, clear solutions are formed at concentrations as high as 20 wt%. The microstructure in these clear solutions has been investigated recently by fluorescence quenching and found to be made up of small spherical micelles.[51] These micelles are apparently quite 'normal' and behave much like their single-chain counterparts; they grow with chain length (see table 9), salt concentration (table 9) and surfactant concentration (figure 9) in ways that are comparable to single-chain micelles.

Similar counterion effects are also seen when the counterions of anionic double-chained surfactants are complexed with macrocyclic compounds like the cryptate.[52] Addition of C222 to SHBS results in the transformation of turbid dispersions of liposomes and other complex structures first into small vesicles and then into micelles as the ratio of C222 to sodium ion is increased.

The fact that such highly hydrophobic molecules can form small micelles seems remarkable at first glance. Micelles are not, however, theoretically forbidden structures for these amphiphiles;[37] if headgroup areas are large enough, local curvature will increase to the point where the spherical micelle criterion, $v/al < \frac{1}{3}$, is met. In the case presently under consideration, the strong hydration of the acetate or hydroxide counterion pulls it out from amongst the headgroups, resulting in the required increase in headgroup repulsions.

The high degree of curvature induced by the acetate or hydroxide ion can be destroyed by a simple acid–base neutralization reaction. For example, when aqueous $2C_{12}N2C_1OAc$ is mixed with equimolar hydrobromic acid, the change of counterion results in the transformation of small, spherical micelles to liposomes and other such structures. The dynamics of this transition have been followed by VEM. In this experiment, the two solutions are pumped at high velocity into a vortex or T-tube type mixer and quickly injected into a microscope flow cell.[50] When the flow is stopped, and the resulting structural transformations are recorded on video tape, the apparatus acts as a 'stopped-flow microscopy' device. As can be seen in figure 10, the micelle-to-liposome transition occurs via worm-like or tubule intermediates. Observation of a similar reaction (between $2C_{12}N2C_1OH$ and HBr) at a sharp interface between the two solutions indicates that the 'worms' are ordered preferentially in the direction of the acid–base concentration gradient.[50] In either case these tubules gradually reassemble into large spherical liposomes until they disappear altogether several hours after the start of the reaction. There is evidence that other structural transformations of double-chained amphiphiles also involve worm-like intermediates.

Temperature effects can also be more dramatic with double-chain surfactants. A typical phase diagram (that of $2C_{10}N2C_1Br$) is shown in figure 11(*a*).[53] The points A (25 °C) and B (75 °C) lie along a line of constant concentration. The microstructure in the two phase region (e.g. point A) consists of the typical stew of aggregate sizes and shapes illustrated in figure 8. The microstructure in the isotropic phase of double-chained surfactants (e.g. point B) has not been extensively investigated because for most surfactants it is accessible only at high temperatures. Recent fluorescence

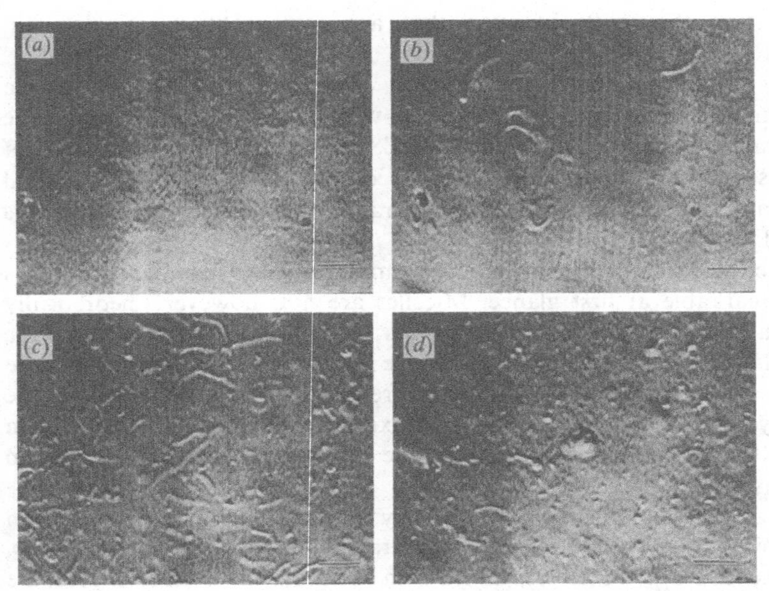

Figure 10. Video micrographs of the dynamic transformation of $2C_{12}N2C_1OAc$ micelles to bromide ($2C_{12}N2C_1Br$) liposomes. Bar = 10 μm. (*a*) 30 s into reaction. Thin filaments or thread-like aggregates form. The nature of these structures is more easily seen 'live' on the video monitor. (*b*) 1 min later. Threads thicken and elongate into 'worm-like' aggregates. (*c*) After 10 min, the entire sample cell is filled with 'worms.' (*d*) 4 h after the initial mixing, most worms have transformed into small spherical particles which slowly grow.

quenching experiments of the aggregation behavior of $2C_{10}N2C_1Br$ reveal that the isotropic phase is composed of micellar aggregates. Like their single-chain counterparts, these micelles grow with concentration. Here the growth is more pronounced (see figure 9) and probably reflects the fact that the system is approaching the formation of a second (lamellar) phase as the concentration is increased. Similar observations of substantial micelle growth near the isotropic–isotropic + lamellar phase boundary have been made with asymmetrical surfactants [54] (surfactants with one long chain and one short chain) as well as the chloride analogue, $2C_{10}N2C_1Cl$, where experimentally accessible isotropic phases exist at room temperature.

Again we see that an increase in thermal motion causes a decrease in local curvature, and the resulting transformation from liposomes to micelles again is predictable from curvature considerations alone.

Using the fluorescence quenching technique we can track the evolution of these micellar aggregates from their liposomal origins as the temperature is raised.[51] We can illustrate this process with a phase diagram of $2C_{12}N2C_1Br$ (figure 11(*b*)).[53] At room temperature (point A), the surfactant exists entirely in large aggregate form and there are no detectable micellar aggregates.[51] As the temperature is raised, local curvature is increased, and

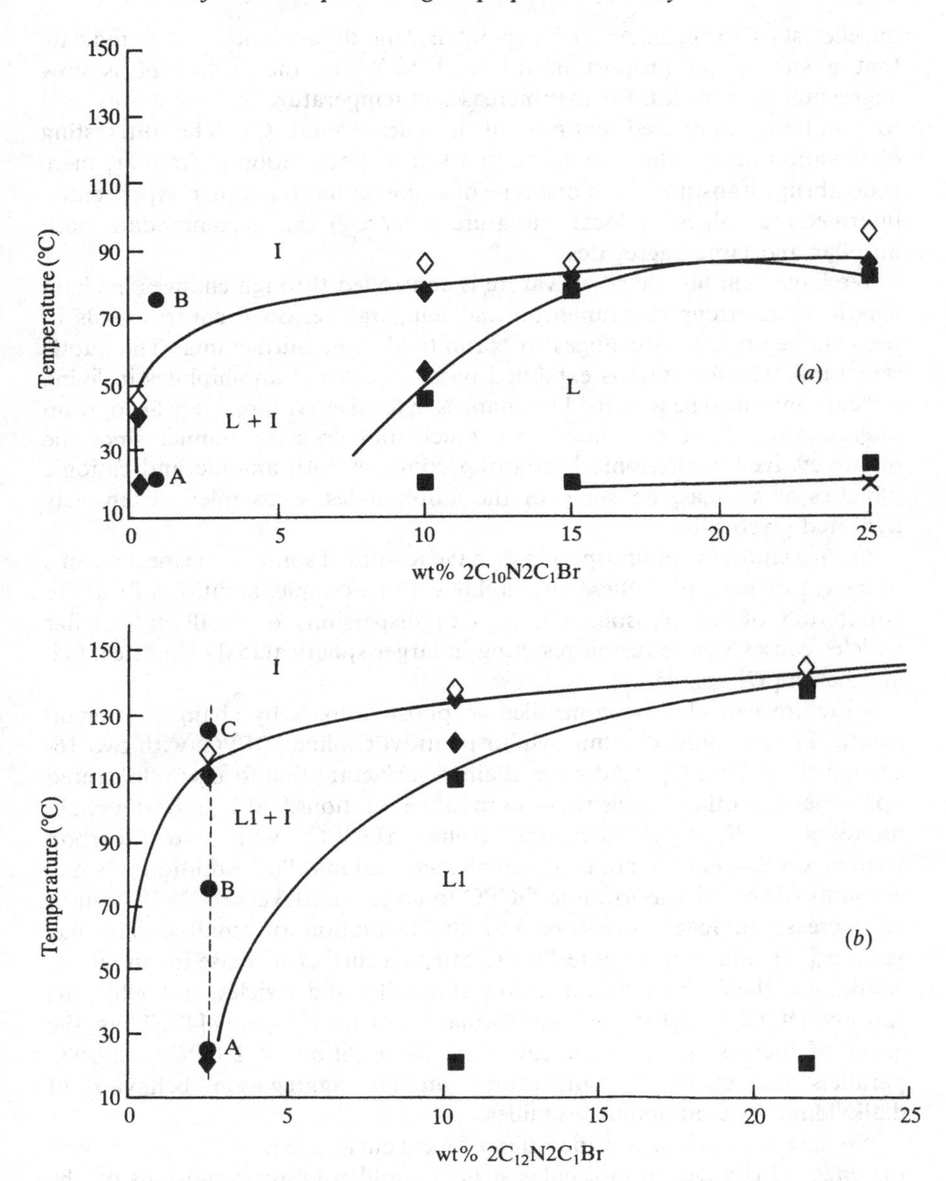

Figure 11. Phase diagrams of dialkyldimethylammonium bromides, adapted from ref. 54. I represents the clear, isotropic phase, L (or L1) is the birefringent lamellar phase while L (or L1) + I is a turbid dispersion containing liposomes and other complicated structures (see figure 8). In both (*a*) ($2C_{10}N2C_1Br$) and (*b*) ($2C_{12}N2C_1Br$), points A, B and C lie along a line of constant concentration.

micelles start to form. At 77 °C (point B), the fluorescence results indicate that a substantial proportion (at least 50%) of the surfactant is now aggregated as micelles. Further increases in temperature will eventually lead to solutions composed entirely of micelles (point C). The interesting observation here is that the liposome-to-micelle transition is gradual; there is no abrupt transition from one type of aggregation to another. Apparently, intermediate values of local curvature ($v/al \approx \frac{1}{2}$) can accommodate both micellar and large aggregates.

We have seen how local curvature is controlled through changes in chain length, ionic strength, counterion and temperature. Such control leads to predictable structural changes in 'primitive' ionic surfactants. The subtle structural transformations exhibited by phospholipid amphiphiles in living systems must also be governed by changes in local curvature. Here headgroup interactions can be controlled in a much more precise manner since the highly evolved zwitterionic headgroup contains both anionic and cationic moieties as well as, on some of the amphiphiles, a complex, extensively hydrated glycocalix.

In vitro studies with phospholipids have identified some of the mechanisms of curvature control in these amphiphiles. For example, addition of Ca^{2+} (a counterion of the phosphate group) to dispersions of small unilamellar vesicles causes vesicle fusion resulting in larger spherical,[53] cylindrical,[56] and helical[57] aggregates.

Structure can also be controlled in phospholipids by changes in chain length. For example, dipalmitoylphosphatidyl choline (DPPC, with two 16-carbon chains) is a typical double-chained surfactant that forms multilayered liposomes and other large structures in saline solutions.[58] The shorter chain homologue, dicaproylphosphatidylcholine (DCPC, with two 6-carbon chains), on the other hand, forms small spherical micelles. Addition of small amounts of the micelle-forming DCPC to large multilayers of DPPC causes an increase in local curvature and the formation of small unilamellar vesicles.[59] Addition of more DCPC causes a further increase in curvature, resulting in the formation of mixtures of micelles and vesicles, and, when the ratio of DCPC to DPPC is high, formation of micelles only.[43] Thus, the effect of increasing the local curvature by addition of DCPC to DPPC parallels the effect of temperature on the aggregation behavior of dialkyldimethylammonium bromides.

We have seen how a consideration of local curvature provides insight into the effect of changes in molecular structure and solution conditions on the aggregation behavior of surfactants. Such considerations were based on the assumption of the isolated aggregate; i.e. the effects of interparticle interactions were not taken into account. In the next section we discuss the nature of such interactions, how they can be measured and how they influence the structural properties of amphiphilic aggregates.

4. Water mediated interactions between amphiphilic aggregates

4.1 *Force measurements*

The role of water in mediating the interactions between solutes remains a major issue in understanding the stability and reactivity of self-organizing systems. Since surfactant aggregates and many biological macromolecules have characteristic dimensions of 10–1000 Å, their interactions fall into the colloid domain. Until recently, much of the information on solvent effects in colloidal systems came from studies on classical sols like gold, silver iodide, titanium oxide, etc. Their stability and coagulation behavior are in accord with the predictions of the classical DLVO theory.[60] This theory draws a clear distinction between repulsive electrostatic forces that keep particles apart and the attractive van der Waals forces which are quantum mechanical in origin and have a power law dependence. The repulsive forces decay exponentially with distance, D, with a range set by concentration, n, and valencies, z, of the intervening electrolyte. In other words they depend on ψ°, the surface potential. The change in potential energy, V_R, with distance, D, is given by

$$V_R = \frac{64\, nkT}{\kappa}\left[\frac{\exp(z\psi^\circ/2kT)+1}{\exp(z\psi^\circ/2kT)-1}\right]\exp(-KD) - \frac{A}{12\pi D^2} \tag{11}$$

where A is the Hamaker constant [61] which is a measure of the van der Waals attraction interactions.

In this theory, we approximate the electrolyte as point charges immersed in a continuum characterized by bulk dielectric constant and viscosity. Why this approximation works so well has posed a considerable puzzle, since the particle surfaces move together to distances where size effects cannot be ignored. Recent theoretical studies by Wennerström and coworkers [62] and by Kjellander and Marcelja [63] have established that two additional interactions exist – a finite ion size effect, which adds an additional repulsive term, and a correlated ionic interaction, which adds an additional attractive term. In most situations these effects cancel one another and produce the curve predicted by the continuum DLVO theory.

In colloid science the DLVO theory plays a role analogous to the Debye–Hückel theory for simple electrolytes. It is the guiding intellectual beacon. In spite of its scientific triumphs, there are significant discrepancies. For example, the DLVO theory has not dealt successfully with the properties of colloids consisting of hydrocarbon moieties (amphiphilic aggregates) in water, with the stability and coalescence of gas bubbles in salt solution, with clay swelling, or with the observed spacing of multilayers. In these systems the attractive interactions are considerably smaller and additional short-range forces which depend on discrete solvent properties become important. Understanding interactions in amphiphilic systems thus depends on understanding solvent mediated forces that go beyond the DLVO theory.

The recent advances in our ability to measure interaction forces directly has provided new insights. There are four main experimental techniques for direct force measurements:

(1) The oldest and most limited involves pressure measurements across soap films as a function of distance.[64]

(2) A much more versatile technique involves setting the osmotic pressure (i.e. the force per unit area) by fixing the activity of water, and measuring the resulting separation between the microstructures by x-ray spectroscopy.[65] Using this technique, Parsegian and coworkers established the existence of short-range hydration forces between phospholipid bilayers. This short-range repulsive force plays a major role in setting the 17–20 Å equilibrium spacing observed in these bilayer systems.[66] More recently, they have applied this technique to DNA.[67]

(3) The third technique employs the surface forces apparatus (SFA) which will be described in detail below.

(4) Recently, it has been reported that a modified version of the scanning tunneling microscope [68] can be used to measure forces on non-conducting surfaces immersed in water directly. With heterogeneous surfaces such force measurements will permit resolution of nanometer dimensions. While this technique is still in its infancy, it is very promising.

The SFA was originally developed by Tabor and Winterston [69] and subsequently improved by Israelachvili and Adams.[70] The SFA permits forces to be measured directly over a distance range of 0–10000 Å with a resolution of ± 1 Å. In order to obtain this distance resolution, however, the surfaces must be molecularly smooth. To date, mica provides the only successful base surface. Mica surfaces are mounted onto glass lenses which are ground as cylindrical surfaces and are brought together in a crossed cylindrical geometry. The forces are measured by the deflection of a leaf spring to which one of the mica surfaces is attached. The separation between the two mica surfaces is determined by transmitting a beam of white light through the mica and observing the interference fringes [71] formed as a result of multiple reflections between the silvered backs of the two surfaces.

Natural mica is negatively charged with one ionic site per 50 $Å^2$; the counterions are potassium. In low dielectric constant solvents mica behaves as an uncharged surface. In high dielectric constant solvents some of the counterions dissociate and the surfaces become charged. In carbon dioxide saturated water, the potassium ions on the mica surface are replaced by protons (relative solution concentrations are $[K^+] \approx 10^{-11}$ M and $[H^+] \approx 10^{-5}$ M). As will be developed in detail below, ion exchange on the mica surface plays an important role in understanding SFA measurements.

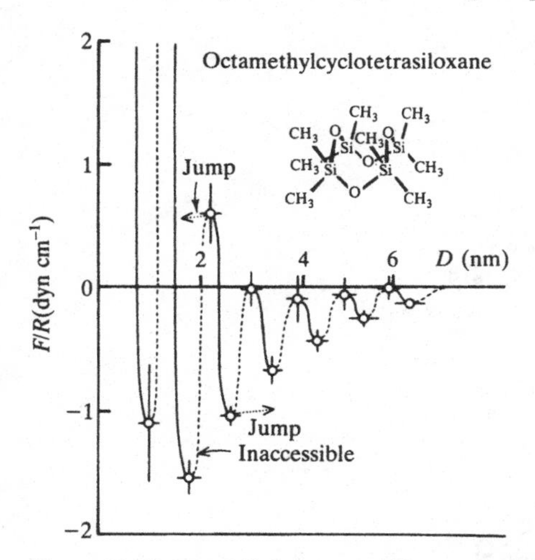

Figure 12. Force vs distance curves for octamethylcyclotetrasiloxane between mica surfaces in the SFA. The oscillations in the curve correspond to successive layers of solvent being squeezed out as the surfaces are brought together. From ref. 72.

4.2. *Nonamphiphilic systems*

An example of solvent induced forces comes from measurements on liquids, composed of spherical molecules, such as cyclohexane or octamethylcyclotetrasiloxane.[72] At distances equal to approximately ten times the molecular diameter, periodic oscillations in the force vs distance curves like those shown in figure 12 are observed. The distance between the minima in these oscillations corresponds to the diameter of the solvent molecules and to successive layers of solvent being squeezed out as the surfaces are brought together. No oscillations are detected with linear hydrocarbons like octane since these flexible molecules can rearrange to fill all available space between the mica surfaces. With extraordinarily careful measurements, oscillations in water can also be detected.[73] These are superimposed upon the normal attractive–repulsive DLVO curve usually measured. Oscillatory force curves are obtained in EAN and in EAN–water mixtures.[74]

The simplest experiments in aqueous solution involve measurements with added alkali metal ions.[75] In carbon dioxide saturated water, the measured curves can be fitted to the DLVO equation assuming surface potential of ≈ -130 V, a Debye length corresponding to carbon dioxide saturated water and an attractive interaction calculated using a Hamaker constant for mica surfaces separated by water. In SFA measurements, the mica surfaces jump into adhesive contact when the force between the surfaces exceeds the force exerted by the calibrated leaf spring. In a typical water experiment, this corresponds to a jump distance that occurs when the two mica surfaces reach a separation of 30 Å.

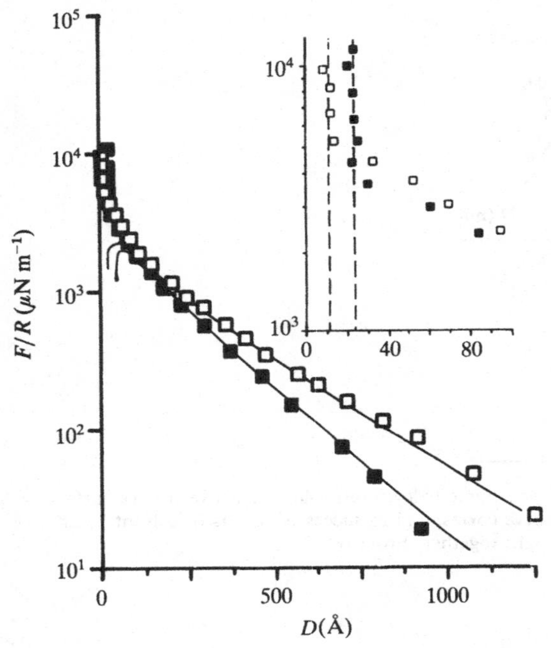

Figure 13. Force between mica surfaces vs distance for solutions of 7.7×10^{-5} M KOH, pH ≐ 10, and with 5.4×10^{-4} M C222 added. (Inset: expanded distance scale force plot showing difference between 'contact' distances with (23 Å) and without (10 Å) cryptate complex.) For potassium hydroxide only the surface potential is fitted to the 115 ± 5 mV, and $\kappa^{-1} = 270$ Å. With added C222 there is a slight decrease to 100 ± 5 mV and $\kappa^{-1} = 198$ Å. Solid lines correspond to best fits to DLVO theory offset by the experimental contact distances.

Upon addition of alkali metal ions, the force curves exhibit a short-range repulsion that prevents the surfaces from jumping into adhesive contact. This repulsive force decays exponentially with a decay constant equal to the diameter of a water molecule. Experiments in which the concentration of ions and pH are varied establish that the force curves can be described in terms of an equilibrium between adsorbed alkali metal ions which give rise to the short-range hydration forces and protons which do not. Ion specificity is manifested by different adsorption constants and ionic site size required to fit the data. There is evidence that, as the mica surfaces approach one another, the relative concentrations of adsorbed protons and other ionic species change. This can be prevented by employing basic solutions. For example, in the presence of 7.7×10^{-5} M KOH (figure 13 open symbols), the surfaces move together, but the distance of closest approach is 10 Å (figure 13, inset), greater than that observed in carbon dioxide saturated water.[48]

Another example of the role of ion exchange in surface force measurements is illustrated in figure 13. As described in the previous section, complexation

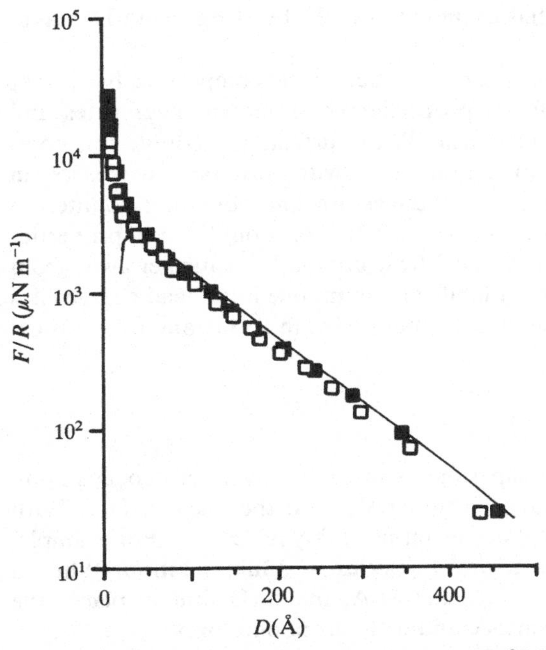

Figure 14. Force vs distance for solutions of 1.0×10^{-3} M KCl, 7×10^{-5} M KOH and same with 1.0×10^{-3} M C222. The force curve for cryptate data is moved to the left by 14 Å, showing similarity of potential and Debye length in these systems (from a second series of K^+ measurements). Solid line is that of DLVO theory with $\psi^\circ = 85$ mV, $\kappa^{-1} = 90$ Å, offset by +10 Å from 'contact.' From ref. 48.

of alkali metal counterions by macrocyclic molecules results in significant decreases in micellar aggregation numbers with single-chain surfactants and transformation of bilayers to vesicles and micelles with double-chain surfactants. Addition of cryptates like C222 to salt solutions in the SFA provides a way to determine the effect of counterion complexation upon the interfacial behavior on mica. This can be viewed as a model for a fixed charge classical colloidal system.[48] When sufficient C222 is added to the 7×10^{-5} M KOH solution, the contact distance increases to 23 Å (figure 14). Addition of 1×10^{-3} M KCl results in a replacement of the adsorbed K:C222 complex and recovery of the 10 Å contact distance. Upon addition of more C222, the 23 Å barrier is recovered. Thus, the large C222:K^+ complex can be exchanged between the solution and the mica surface.

An important result of these experiments can be seen most directly from an experiment in which 1×10^{-3} M C222 is added to a solution containing 1×10^{-3} M KCl and 7×10^{-5} M KOH (figure 14). This permits the two force curves to be directly compared with and without cryptate. The curves are identical except that the complexed solution is displaced by 14 Å. When this difference in distance is taken into account, analysis with the DLVO theory gives exactly the same potential using the same Debye length and Hamaker constant.

Thus, complexation of counterions by macrocyclic compounds has a very different effect on the interfacial properties of surfactant aggregates and classical colloidal particles like mica. With surfactants, counterion complexation changes headgroup interactions with resultant decreases in aggregate size. With the clays, the changes are immobile and counterion complexation has almost no measureable effect on double-layer properties except to displace the position of the diffuse double-layer further away from the solid surface. The pressure of labile and immobile interfacial charge sites gives rise to very different response to counterions in surfactant and classical colloidal systems.

4.3. *Amphiphilic system*

A more sophisticated level of experiments involving the SFA depends upon devising ways to adsorb or deposit materials onto the mica surface. With amphiphiles, one can often exploit the self-assembly properties. For example, $2C_{16}N2C_1OAc$ self-assembles to form vesicles in dilute solution. When a 5×10^{-5} M solution is injected into the SFA, the surfactant displaces the potassium ions on the mica surface and a 'hydrophobic' monolayer forms, as evidenced by a 27 Å increase in the contact distance when the mica surfaces are moved together.[77] Because all the charges on the mica surface are neutralized by adsorption of the cationic surfactant, no repulsive forces can be detected at any distance.

The interaction between hydrocarbon chains and water that leads to amphiphilic assembly involves short-range attractive forces that are in accord with the predictions of van der Waals theory. However, interactions between the large regions of hydrocarbon encountered in hydrophilic monolayers are less well characterized. The self-assembly described above provides a means for directly measuring these forces.

We can grasp the significance of the data [76] most readily by using the Deryagin approximation which relates the interaction energy E per unit area to the measured force ($E = F/2R$). Interaction energy can be fitted to the form, as shown in figure 15, where D is in nanometers. The two curves merge at about 15 nm, demonstrating attractive forces 10–100 times larger than those predicted by continuum theory. When the two hydrophobic surfaces are pulled apart, they jump out to about 100 nm and cavitation results in the formation of a water vapor bubble that is 20 nm in diameter. This indicates the existence of strongly hydrophobic surfaces in contact with water. Measurements by Claesson, Blom, Harder and Ninham confirm these observations.[77] These results are unexpected and raise many unanswered questions.

Increasing the concentration of $2C_{16}N2C_1$ ion results in the self-assembly

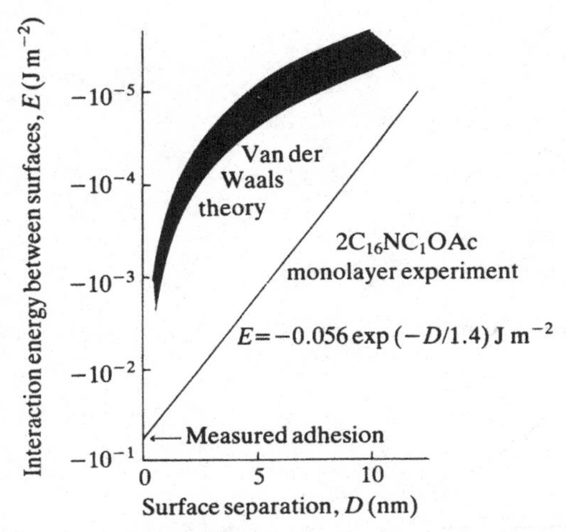

Figure 15. Interaction energy per area (E) for flat plates as a function of their separation distance. Note that the attraction energy at separations of 1–7 nm is 10–100 times larger than predicted by continuum van der Waals theory. Theoretical (shaded) and experimental curves merge at long separations (about 15 nm). From ref. 76.

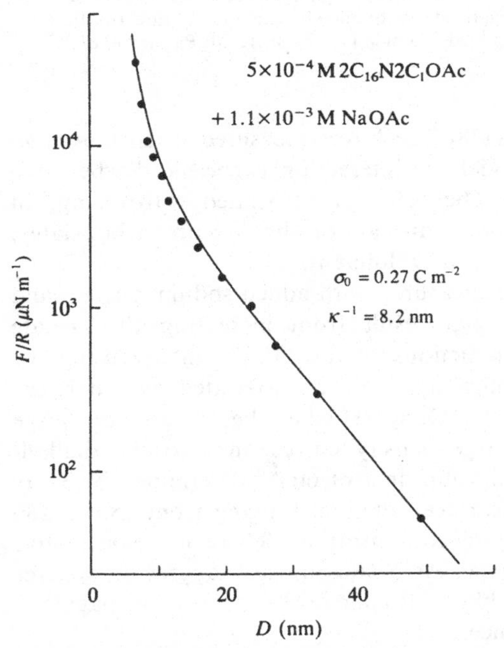

Figure 16. Forces measured between absorbed bilayers of $2C_{16}N2C_1OAc$ in 1.1×10^{-3} M NaOAc. The data agree well with the predictions of DLVO theory using the known Hamaker constant and the Debye length calculated from Debye-Hückel theory. Note the lack of an attractive potential minimum, indicating that acetate aggregates will not coagulate with added acetate ion (compare figure 17). From Evans, D. F. and Ninham, B. W., *J. Phys. Chem.* **90**, 226 (1986).

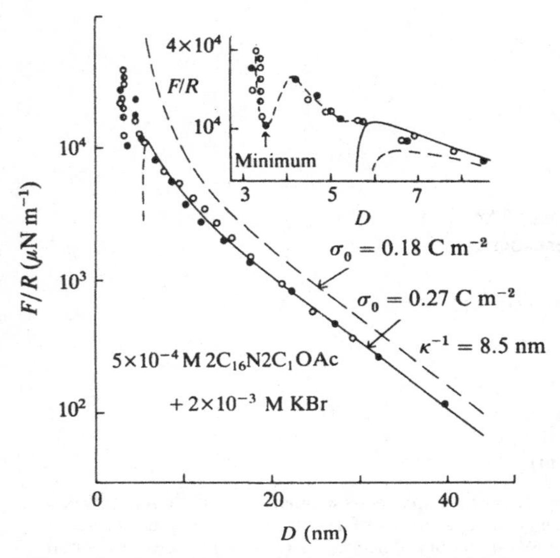

Figure 17. Forces measured between absorbed bilayers of $2C_{16}N2C_1OAc$ acetate in 2×10^{-3} M KBr. The solid and dashed lines give predictions from DLVO theory assuming 7% and fully dissociated bilayers, respectively. Note the attractive potential minimum that is now present, indicating that, unlike acetate aggregates with added acetate ion (figure 16), the bromide aggregates will coagulate with added bromide ion. From R. M. Pashley *et al.*, *J. Phys. Chem.* **90**, 1637 (1986).

of bilayers on the mica surfaces.[78] The forces measured when these two bilayers are moved together model the interaction experienced when two vesicles approach one another. The behavior associated with change in counterions (particularly halide vs hydroxide or short-chain carboxylates) described in the previous section is very different.

Figures 16 and 17 show forces measured with added sodium acetate and potassium bromide. The data that result from increasing the acetate concentration agree with the predictions of the DLVO theory using the known Hamaker constant for mica–hydrocarbon separated by water, the Debye length calculated from the Debye–Hückel theory, and a charge density of $0.27\ C\,m^{-2}$. This charge density corresponds to the dialkyldimethylammonium bilayer headgroup area of 60 $Å^2$ determined by x-ray crystallographic analysis and a surface potential ranging from 280 to 180 mV, depending on sodium acetate concentration. There is no attractive potential minimum at any distance. As a consequence, aggregate systems with highly hydrated counterions like hydroxide or acetate will not coagulate, but remain repulsive at all distances.

Very different behavior results when we add potassium bromide (figure 17). The surface charge density is considerably lower ($0.03\ Cm^{-2}$ at 1×10^{-4} M KBr) and it decreases when we increase the salt concentration. A

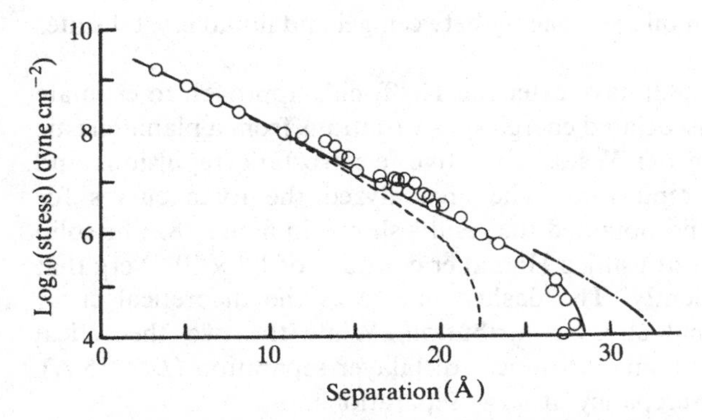

Figure 18. Optimum fit of the theoretical total stress (full lines) to data derived from electrically neutral egg phosphatidylcholine experiments (○); the van der Waals attraction coefficient A_H was found to be 1.1×10^{-13} erg (1 erg = 10^{-7} J) to agree with the observed separation distance at full hydration *and* with adhesion energy values derived from experiments on giant bilayer vesicle adhesion ($\approx 10^{-2}$ erg cm^{-2}). For comparison, a van der Waals attraction coefficient comparable to earlier estimates (7×10^{-14} erg) was used to calculate the total stress shown as the chain line. Plotted as the dashed line is the elastic planar stress associated with the optimum fit (minus the contribution of fluctuation stress); this is the level of stress expected to act between immobilized bilayers. The hydration repulsion is characterized by a magnitude $P_{hyd} = 9 \times 10^9$ dyne cm^{-2} (1 dyne = 10^{-5} N) and a decay length of 2.2 Å.

secondary minimum develops at sufficiently high concentrations of potassium bromide, and the two bilayers each slough off their outermost layer upon approaching one another so that the remaining monolayers come into hydrophobic contact. When the surfaces are separated, the bilayers reform. As a result of low potentials (~ 70 mV) associated with the high degree of counterion binding, surfactant aggregates will increase in size and coagulate under the influence of the van der Waals attractive forces.

Similar counterion specificity is seen in formamide.[79] The force curves for acetate give complete agreement with the DLVO theory if we assume no counterion binding, while those for bromide require 80% counterion binding. This difference simply reflects the fact that solvation effects associated with specific counterions located near charged interfaces must be due to short-range interaction arising from hydrogen bonding and dipolar effects and consequently involve only a few solvent molecules.

A major question in understanding the interaction of membranes and liposomes concerns the role of longitudinal thermal mechanical fluctuations in determining bilayer equilibrium separations. In 1978, Helffrich [80] proposed that such fluctuations could contribute to the repulsive forces that stabilize bilayers. He expressed these fluctuations in terms of undulating displacements of the bilayers from a planar configuration. In the original formulation, the magnitude and consequent steric repulsion in the absence of long-range interbilayer forces was computed. Sornette and Ostrowsky [81] interpreted the change in bilayer spacing between gel and liquid crystal states in terms of these ideas.

Evans and Parsegian [82] have extended Helffrich's approach to estimate fluctuations (and their associated energies) as variations from a planar elastic state which include van der Waals attractive, electrostatic repulsions and short-range hydration repulsions. They reanalyzed the force curves for neutral phospholipids and obtained the results shown in figure 18. The solid line gives the theoretical fit using a Hamaker constant of 1.1×10^{-13} erg that was obtained independently. The dashed line gives the theoretical curve without the fluctuational stress contribution. While the two theoretical curves almost coincide at small distances of bilayer separation ($D \leqslant 15$ Å), there is considerable discrepancy at large separations.

Further support for the role of fluctuations in determining the long-range behavior of amphiphilic bilayers comes from a comparison [83] of osmotic pressures and SFA measurements.[84] This permits a direct comparison of the system with fluctuations and where they have been suppressed by adsorption onto mica. The SFA curves agree with the theory without fluctuations while those for the osmotic pressure agree with the curve with fluctuation included. Furthermore, below the chain melting temperature where Helffrich fluctuations are greatly diminished, the force curves from the SFA and osmotic pressure measurements on phospholipids are consistent.

5. Conclusion

Water remains as ethereal as ever. After decades of intense study, we seem to know more about what it isn't than what it is. The rigorous debate about whether water is best described by a one-state or a two-state model has diminished. In the heat of that battle, Henry Frank pointed out the four possible conclusions – one side was right – the other wrong, or vice versa, or both were wrong or both were right. He argued for the first option, instinctively leaned towards the third and dismissed the fourth on logical grounds. In fact, both approaches were simultaneously right and wrong in the sense that they each glimpsed only part of water's complexity. Similarly, the effect of water structure on the self-assembly of schizophrenic amphiphilic molecules has been a source of considerable confusion. Self-contradictory inferences based on thermodynamic data can be reconciled when freed from the rigor mortis of crystallography, fluctuations are given their due and the role of compensated behavior recognized. Understanding water and how it guides and controls amphiphilic and macromolecular processes remains a major scientific and intellectual challenge. The hypotheses we formulate become more sophisticated, but the subtle and facile nature of water remains an enigma.

References

1. J. E. Gordon, *The Organic Chemistry of Electrolyte Solution*, Wiley, New York, 1975, pp. 158–62.
2. M. Ramadan, D. F. Evans & R. Lumry, *J. Phys. Chem.* **87**, 4538 (1983).
3. M. Ramadan, D. F. Evans, R. Lumry & S. Philison, S. *J. Phys. Chem.* **89**, 3405 (1985).
4. D. F. Evans, A. Yamauchi, R. Roman & E. Z. Casassa, *J. Colloid & Interf. Sci.* **86**, 89 (1982).
5. D. F. Evans, A. Yamauchi, G. J. Wei & V. A. Bloomfield, *J. Phys. Chem.* **87**, 3537 (1983).
6. A. Ray, *Nature* **231**, 313 (1971); *J. Am. Chem. Soc.* **91**, 6511 (1969).
7. R. J. Lemire and P. G. Sears, *J. Chem. Eng. Data*, **22**, 376 (1977).
8. R. G. Laughlin & R. L. Munyan, *J. Phys. Chem.* **91**, 3299 (1987).
9. A. H. Beesley, D. F. Evans & R. G. Laughlin, *J. Phys. Chem.* **92**, 791 (1988).
10. D. F. Evans, D. J. Mitchell & B. W. Ninham, *J. Phys. Chem.* **88**, 6344 (1984).
11. D. F. Evans & B. W. Ninham, *J. Phys. Chem.* **87**, 5025 (1983).
12. R. Lumry, E. Battistel & C. Jolicoeur, *J. Chem. Soc., Faraday 1982*; R. Lumry & H. S. Frank *Proc. 6th Int. Biophys. Congr.* **7**, 544. (1978).
13. D. F. Evans, S.-H. Chen, G. W. Schriver & E. M. Arnett, *J. Amer. Chem. Soc.* **103** (2), 481 (1981).
14. D. Mirejovsky & E. M. Arnett, *J. Am. Chem. Soc.* **105**, 112 (1983).
15. D. F. Evans, E. W. Kaler & W. J. Benton, *J. Phys. Chem.* **87**, 533 (1983).
16. W. Tamura-Lis, L. J. Lis & P. J. Quinn, *J. Phys. Chem.* **91**, 4625 (1987).
17. T. J. O'Leary & J. W. Levin, *J. Phys. Chem.* **88**, 4074 (1984).
18. M. Allen, D. F. Evans & R. Lumry, *J. Solution Chem.* **14**, 549 (1985).
19. D. K. Magnuson, J. W. Bodley & D. F. Evans, *J. Solution Chem.* **13**, 583 (1984).
20. I. Rico & A. Lattes, *J. Phys. Chem.* **90**, 5870 (1986).
21. A. Belmajdoub, J. P. Marchal, C. Canet, I. Rico & A. Lattes, *New J. Chem.* **11**, 415 (1987).
22. D. W. Larsen, S. B. Rananavare & S. E. Friberg, *J. Amer. Chem. Soc.* **106**, 1848 (1984).
23. K. Shinoda, *J. Phys. Chem.* **81**, 1300 (1977).
24. H. S. Frank & M. W. Evans, *J. Chem. Phys.* **13**, 507 (1945).
25. D. W. Davidson, in *Water – A Comprehensive Treatise*, Vol. 2, Clathrate Hydrates (ed. F. Franks) Plenum, New York, 1973, chapter 3.
26. W. Kauzmann, *Adv. Protein Chem.* **14**, 1 (1959).
27. D. F. Evans & P. J. Wightman, *J. Colloid Interf. Sci.* **86**, 515 (1982).
28. A. H. Beesley, *Ph.D. Thesis*, Department of Chemical Engineering and Materials Science, University of Minnesota, 1987.
29. R. Lumry, & R. B. Gregory, in *The Fluctuating Enzyme*, (ed. G. R. Welch) Wiley-Interscience, New York, 1986.
30. T. H. Benzinger, *Nature* (*London*) **229**, 100 (1971).
31. T. H. Benzinger & C. Hammer, *Curr. Top. Cell. Regul.* **18**, 475 (1981).
32. R. Lumry and H. S. Frank, *Proc. 6th Int. Biophys. Congr.* **7**, 544 (1978).
33. J. C. Slater, *Introduction to Chemical Physics*, McGraw-Hill, New York, 1939, chapter 13.

34. H. S. Frank, private communication.
35. Data for these calculations taken from K. Pitzer and L. Brewer, *Thermodynamics*, McGraw-Hill, London, 1961.
36. C. Tanford, *The Hydrophobic Effect*, 2nd edition, John Wiley & Sons, New York, 1980.
37. D. J. Mitchell & B. W. Ninham, *J. Chem. Soc. Faraday Trans. 2* **77**, 601 (1981).
38. B. Jönsson & H. Wennerström, *J. Phys. Chem.* **91**, 338 (1987).
39. A. Ben-Shaul, J. Szleifer & W. M. Gelbart, *Proc. Natl. Acad. Sci. U.S.A.* **81**, 4601 (1984).
40. J. N. Israelachvili, D. J. Mitchell & B. W. Ninham, *J. Chem. Soc., Faraday Trans.* 2 **72**, 1525 (1976).
41. T. Imae, R. Kamiya & S. Ikeda, *J. Colloid Interf. Sci.* **108**, 215 (1985).
42. E. Roelants & F. C. DeSchryver, *Langmuir* **3**, 209 (1987).
43. D. D. Miller, L. M. Magid & D. F. Evans, *J. Phys. Chem.*, **93**, 323 (1989).
44. J. E. Brady, D. F. Evans, G. G. Warr, F. Grieser & B. W. Ninham, *J. Phys. Chem.* **90**, 1853 (1986).
45. L. Sepúlveda & J. Cortés, *J. Phys. Chem.* **89**, 5322 (1985).
46. P. Lianos & R. Zana, *J. Phys. Chem.* **87**, 1289 (1983).
47. D. F. Evans & B. W. Ninham, *Phys. Chem.* **87**, 5025 (1983).
48. D. F. Evans, J. B. Evans, R. Sen & G. G. Warr, *J. Phys. Chem.* **92**, 784 (1988).
49. G. G. Warr & D. F. Evans, *Langmuir*, **4**, 217 (1988).
50. D. D. Miller, J. R. Bellare, D. F. Evans, Y. Talmon & B. W. Ninham, *J. Phys. Chem.* **91**, 674 (1987).
51. D. D. Miller & D. F. Evans, *J. Phys. Chem.*, in press.
52. D. D. Miller, D. F. Evans, G. G. Warr & J. R. Bellare, *J. Colloid Interf. Sci.* **116**, 598 (1987).
53. G. G. Warr, R. Sen, D. F. Evans & J. E. Trend, *J. Phys. Chem.*, **92**, 774 (1988).
54. P. Lianos, J. Lang & R. Zana, *J. Colloid Interf. Sci.* **91**, 276 (1983).
55. J. Wilschut & D. Papahadjopoulos, *Nature* **281**, 690 (1979).
56. D. Papahadjopoulos, W. J. Vail, K. Jacobson & G. Poste, *Biochim. Biophys. Acta* **394**, 483 (1975).
57. K.-C. Lin, R. M. Weis & H. M. McConnell, *Nature* **296**, 164 (1982).
58. T.-L. Lin, S.-H. Chen, N. E. Gabriel & M. F. Roberts, *J. Amer. Chem. Soc.* **108**, 3499 (1986).
59. N. E. Gabriel & M. F. Roberts, *Biochemistry* **25**, 2812 (1986).
60. B. V. Derjaguin, & L. D. Landau, *Acta Phys. Chim. USSR* **14**, 633 (1941); E. J. W. Verwey & J. Th. G. Overbeek, Theory of Stability of Lyophobic Colloids, Elsevier, Amsterdam, 1948.
61. D. B. Hough & L. R. White, *Adv. Colloid Interf. Sci.* **14**, 3 (1980).
62. L. Goldbrand, B. Jönsson, H. Wennerström & P. Linse, *J. Chem. Phys.* **80**, 2221 (1984); A. Khan, B. Jönsson, & H. Wennerström, *J. Chem. Phys.* **89**, 5180 (1985).
63. R. Kjellander & S. Marcelja, *J. Phys. Chem.* **90**, 1230 (1986).
64. J. Lyklema & K. J. Mysels, *J. Am. Chem. Soc.* **87**, 2539 (1965).
65. D. M. LeNeveu, R. P. Rand & V. A. Parsegian, *Nature*, **259**, 601 (1976).

66. L. J. Lis, M. McAlister, N. Fuller, R. P. Rand & V. A. Parsegian, *Biophys. J.* **37**, 661 (1982).
67. D. C. Rau, B. Lee & V. A. Parsegian, *Proc. Natl. Acad. Sci., USA* **81**, 2621 (1984).
68. B. Drake, R. Sonnefield, J. Schneir & P. K. Hansma, *Rev. Sci. Instrum.* **57**, 441 (1986).
69. D. Tabor & R. H. S. Winterston, *Proc. Roy. Soc.* **A312**, 435 (1969).
70. J. N. Israelachvili & G. E. Adams, *J. Chem. Soc. Faraday Trans I.* **74**, 975 (1978).
71. S. Tolansky, *Multiple-Beam Interferometry of Surfaces and Films*, Oxford University Press (Clarendon) London, 1949.
72. H. K. Christenson, R. G. Horn & J. N. Israelachvili, *J. Colloid Interf. Sci.* **88**, 79 (1983).
73. J. N. Israelachvili & R. M. Pashley, *Nature* **306**, 249 (1983).
74. R. G. Horn, D. F. Evans & B. W. Ninham, *J. Phys. Chem.* **92**, 3531 (1988).
75. R. M. Pashley, *J. Colloid Interf. Sci.* **83**, 531 (1981).
76. R. M. Pashley, P. M. McGuiggan, B. W. Ninham, & D. F. Evans, *Science* **221**, 1047 (1983).
77. P. M. Claesson, G. E. Blom, P. C. Harder & B. W. Ninham, *J. Colloid Interf. Sci.* **114**, 234 (1986).
78. R. M. Pashley, P. M. McGuiggan, B. W. Ninham, J. Brady & D. F. Evans, *J. Phys. Chem.* **90**, 1637 (1986).
79. J. B. Evans & D. F. Evans, *J. Phys. Chem.* **91**, 3820 (1987).
80. W. Helffrich, *Z. Naturforsch.* **339**, 305 (1978).
81. D. Sornette & N. Ostrowsky, *J. Phys.* (Paris) **45**, 265 (1984).
82. E. A. Evans & V. A. Parsegian, *Proc. Natl. Acad. Sci. USA* **83**, 7132 (1986).
83. A. Parsegian, private communication.
84. J. Mawa & J. N. Israelachvili, *Biochem.* **24**, 4608 (1985).

Solution properties of water-soluble polymers

D. EAGLAND

School of Pharmaceutical Chemistry, University of Bradford, Bradford BD7 1DP

1. Introduction

Water-soluble polymers are widely used in a variety of roles, they are valuable for their 'thickening' capabilities in, for example, emulsion paint formulation, and for their adsorption capability in such industrial contexts as emulsion polymerisation. Certain water-soluble polymers, because of their exceptional biocompatibility are used increasingly in the food and drug industries as extenders and prolonged release agents. The increasing concern expressed with regard to contamination of the environment by organic solvent based processes, means that in the future increasing reliance will be placed upon materials which are soluble in water.

Water-soluble polymers in aqueous solution in many ways do not behave in a manner comparable with polymers dissolved in organic solvents, due to the uniqueness of the solvent – water. The high capability of water for hydrogen bonding, its high dielectric constant and the capacity for hydrophobic hydration of non-polar groups, mean that solution behaviour of water-soluble polymers can be quite startlingly different to that of polymers dissolved in organic solvents, one illustrative example being a tendency to precipitate out of solution upon warming!

It is not possible, within the compass of this review, to examine exhaustively the solution behaviour of every kind of water-soluble polymer, it is rather, the intention to highlight the areas in which uniqueness is exhibited, and those where contention as to the explanation for the behaviour exists.

It seems that the most logical point to start is by defining the term 'polymer', as used in this review – broadly speaking, molecules of molar mass greater than approximately 1000, molecules of lower mass being known as 'oligomers', however the distinction between them is very loose. The polymer may consist of repeats of the same unit, a 'homopolymer', or repeats of different monomer units, a 'copolymer' – the distribution of the monomer units within the macromolecular chain leading to a 'random' copolymer, or to a, more or less, 'block' copolymer.

Polymers of the type

—(CH—CRR′)—

known as vinyl or acrylic polymers have an important structural feature, the different possible steric configurations of the substituted carbon atom, *viz* see diagram 1. Should all the repeat units be of the same type, say type (*a*) then the chain is 'isotactic', alternate (*a*) and (*b*) is 'syndiotactic', whilst random (*a*) and (*b*) is termed 'atactic'.

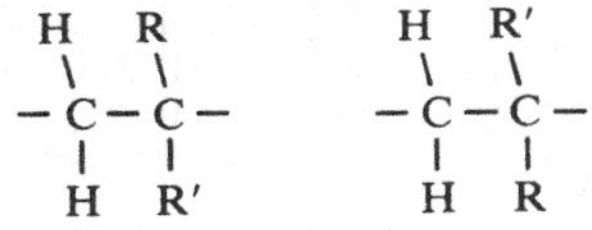

```
 H   R           H   R'
  \   \           \   \
 -C - C-         -C - C-
  |   |           |   |
  H   R'          H   R

   (a)              (b)
```

Diagram 1

The number of repeat units within the macromolecule is termed the 'degree of polymerisation' (DP), of the polymer; it is an inevitable consequence of the process of manufacture that a polymer sample is likely to contain a range of DPs.

To summarise, a polymer sample may contain more than one type of monomer, which may be distributed in a variety of ways within the molecule, and additionally, may be oriented in different ways with respect to nearest neighbour similar monomer units; further, the sample may contain macromolecules with a wide variety of DPs, and if the conditions of polymerisation are severe – high levels of initiator and high degrees of polymerisation – then the macromolecule is likely to contain a significant amount of chain branching. Low levels of initiator and low polymerisation yields are more likely to give unbranched macromolecules. Since each one of these factors can have an effect on the conformation and solubility of the polymer, it is easy to see that considerable variations are possible.

Thus far the reasons for the polymer being soluble in water have not been considered – of necessity hydrophilic functional groups must be present in the macromolecule, groups such as —OH, —COOH, —$CONH_2$, or ionic groups such as —COO^-, —SO_3H^-, —SO_4^{2-}, —NH_3^+. Such groups may be contained in the monomer prior to polymerisation, or introduced subsequently by chemical modification.

Water-soluble polymers fall conveniently into classification according to origin (see appendix for glossary of polymer names):

(i) natural polymers (biopolymers), typified by proteins, nucleic acids and polysaccharides

(ii) chemically modified natural polymers; for example HEC, CMC, HES – may also be termed semi-synthetic polymers.

(iii) synthetic polymers; PVP, POE/PEG, PAAm.

Table 1. *Important water-soluble polymers*

Group	Name	Acronym	Monomer unit
Vinyl polymers	Polyethylene–sulphonic acid	PESA	$-[CH(SO_3H)-CH_2]-$
	Polystyrene–sulphonic acid	PSSA	$-[CH(C_6H_4-SO_3H)-CH_2]-$
	Polyvinyl alcohol	PV–OH (PVAL, PVA1)	$-[CH(OH)-CH_2]-$
	Polyvinylamine	PVAm	$-[CH(NH_2)-CH_2]-$
	Polyvinylmethoxy–acetal	PVMA	$-[CH-CH_2-CH_2-CH_2]-$ with O–CH(CH$_2$–O–CH$_3$)–O bridge
	Polyvinyl methyl ether	PVME	$-[CH(O-CH_3)-CH_2]-$
	Polyvinylmethyl–oxazolidone	PVMO	$-[CH-CH_2]-$ with N-linked methyl-oxazolidone ring (CH_3, O, C=O)
	Polyvinyl–pyrrolidone	PVP	$-[CH-CH_2]-$ with N-linked pyrrolidone ring (C=O)
Vinyl polymers (cont.)	Poly-4-vinyl pyridine	P4VP	$-[CH-CH_2]-$ with 4-pyridyl ring (N)
	Poly-4-vinyl–pyridine–N–oxide	P4VPO	$-[CH-CH_2]-$ with 4-pyridyl ring ($N^{\oplus}-O^{\ominus}$)
	Poly–4–vinyl–N–alkyl–pyridinium salts	P4VRPX	$-[CH-CH_2]-$ with 4-pyridinium ring ($N^{\oplus}-R$, $X^{\ominus}$)
	Polyvinylsulphuric acid	PVSA	$-[CH(O-SO_3H)-CH_2]-$
Polyimines and oxides	Polyethylene imine	PEI	$-[NH-CH_2-CH_2]-$
	Polyethylene oxide (polyethylene glycol, polyoxyethylene)	PEO	$-[O-CH_2-CH_2]-$
	Polypropylene oxide (polypropylene glycol)	PPO (PPG)	$-[O-C(CH_3)-CH_2]-$
Acrylic polymers	Polyacrylic acid	PAA	$-[CH(COOH)-CH_2]-$
	Polyacrylamide	PAAm	$-[CH(CO-NH_2)-CH_2]-$
	Poly-(N,N–dimethyl–acrylamide)	PDMAAm	$-[CH(CO-N(CH_3)_2)-CH_2]-$
	Poly–(N-isopropyl–acrylamide)	PIPAAm	$-[CH(CO-NHC_3H_7)-CH_2]-$
	Polymethacrylic acid	PMAA	$-[C(CH_3)(COOH)-CH_2]-$
	Polymethacrylamide	PMAAm	$-[C(CH_3)(CO-NH_2)-CH_2]-$
	Polycrotonic acid	PCA	$-[CH(COOH)-CH(CH_3)]-$
	Polyethacrylic acid	PEA	$-[C(C_2H_5)(COOH)-CH_2]-$

Table 2. *Some water-sensitive, marginally water-soluble and associated synthetic homopolymers*

Group	Name	Acronym	Monomer unit
Acrylic Polymers	Polymethyl acrylate	PMA	$\left[CH(CO{-}OCH_3){-}CH_2 \right]$
	Polymethyl methacrylate	PMMA	$\left[C(CH_3)(CO{-}OCH_3){-}CH_2 \right]$
	Poly(2-hydroxyethyl methacrylate) (Poly(ethylene glycol monomethacrylate))	PHEMA (PolyHEMA)	$\left[C(CH_3)(CO{-}O(CH_2)_2{-}OH){-}CH_2 \right]$
Polyoxides	Polyoxymethylene (Polyformaldehyde, Polyacetal)	POM	$\left[O{-}CH_2 \right]$
	Poly(trimethylene oxide) (Polyoxacyclobutane)	P3MO	$\left[O{-}(CH_2)_3 \right]$
	Polyacetaldehyde	PAC	$\left[O{-}CH(CH_3) \right]$
Vinyl Polymers	Poly(vinyl ethyl ether)	PVEE	$\left[CH(O{-}C_2H_5){-}CH_2 \right]$
	Poly(vinyl acetate)	PVAC (PVAc, PV-OAc)	$\left[CH(O{-}COCH_3){-}CH_2 \right]$
	Poly(vinyl acetal)s e.g. formal (R = H) PVBu butyral(R = Bu)	PVFo	$\left[CH{-}CH_2{-}CH{-}CH_2 \right]$ (O—CH(R)—O bridging the two CH)

This review will consider mainly the aqueous solution behaviour of groups (ii) and (iii), the semi-synthetic and synthetic water-soluble polymers.

Table 1 illustrates the structural details of the more common synthetic homopolymers, showing the features responsible for water solubility; table 2 lists closely related homopolymers, which are either marginally water-soluble or only water-sensitive, showing that only a minimum change in molecular constitution may cause a polymer to become water insoluble. Addition of $—CH_2—$ or $—CH_3$ frequently leads to insolubility (PAA to PMA, PEO to P3MO, PMAA to PMMA, PVME to PVEE), or to much reduced water solubility (PEO to PPO); in other cases however, aqueous

Table 3. *Water-soluble copolymers*

AAm/AA	Acrylamide/acrylic acid
EO/PO	Ethylene oxide/propylene oxide
MA/AA	Maleic acid/acrylic acid
MA/VAE	Maleic acid/vinyl alkyl ethers
MAAm/MAA	Methacrylamide/methacrylic acid
S/MA	Styrene/maleic acid
S/SSA	Styrene/styrene sulphonic acid
S/VP	Styrene/vinyl pyrollidone
VOH/VA	Vinyl alcohol/vinyl acetate
VA/VP	Vinyl acetate/vinyl pyrrolidone
VA/VSA	Vinyl acetate/vinyl sulphuric acid

solubility may be retained (PAA to PMAA, PAAm to PMAAm, PVOH to PVME, PESA to PPSA). A single case is known (PEO to PMO), where loss of a methylene group leads to loss of water solubility.

Table 3 indicates some of the more common hydrophilic copolymers, some are produced by direct copolymerisation (EO/PO, MA/AA, MAAm/MAA, S/VP, VA/VP, VA/VSA), others by partial chemical modification of the parent homopolymer (VOH/VA, S/SSA, MA/MMA, MAAm/MAA, AAm/AA). Many samples of so-called homopolymers are often copolymers, produced by incomplete modification of a homopolymer, for example, PVOH and PAAm can have a small but significant copolymeric character because of the method of production; PVOH may contain 1–2% of residual acetate groups, PAAm can contain traces of PAA. The effect of small quantities of the residual polymer upon solution behaviour is particularly well illustrated in figure 1 – depicting the solubility of PVOH as a function of residual acetate content.

The manner in which chemical modification, producing the copolymer, is carried out can lead to significant differences in the solution behaviour; figure 2 illustrates the difference of partial molal volume of several PVOH/VA samples as a function of temperature. Two of the samples (2III and 3II) are of identical DPs and residual acetate content, but were hydrolysed under different conditions, resulting in a much more 'blocky' distribution of acetate groups in one sample when compared to the other[1]. This example illustrates a general problem concerning solution behaviour of copolymers, distribution of the 'blocks' is often as significant for solution behaviour as the extent of the 'blocks' – a factor which has unfortunately very often been neglected in the characterisation of water-soluble polymers.

Some of the most common examples of water-soluble semi-synthetic polymers are derived from cellulose, which is itself water insoluble and must be treated, usually by sodium hydroxide based catalysis reactions, before water-soluble derivatives can be obtained. Cellulose is based upon repeat glucopyranosyl units, which if joined via 1–4 equatorial bonds, β-linkages,

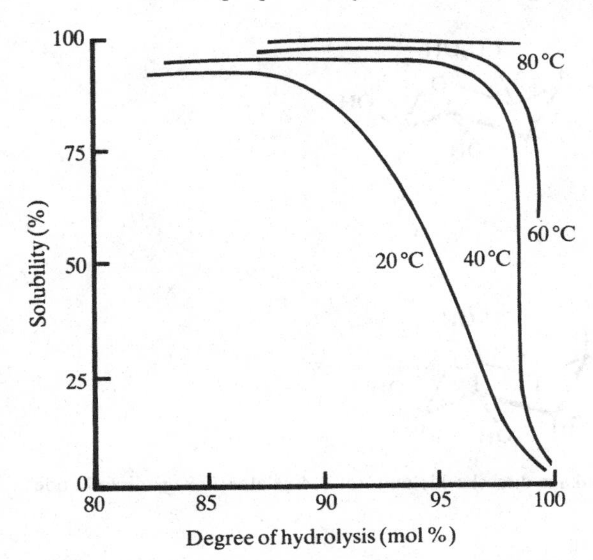

Figure 1. The solubility of PVOH/VA copolymer as a function of residual VA content (polymer DP = 1750) (data from ref. 148).

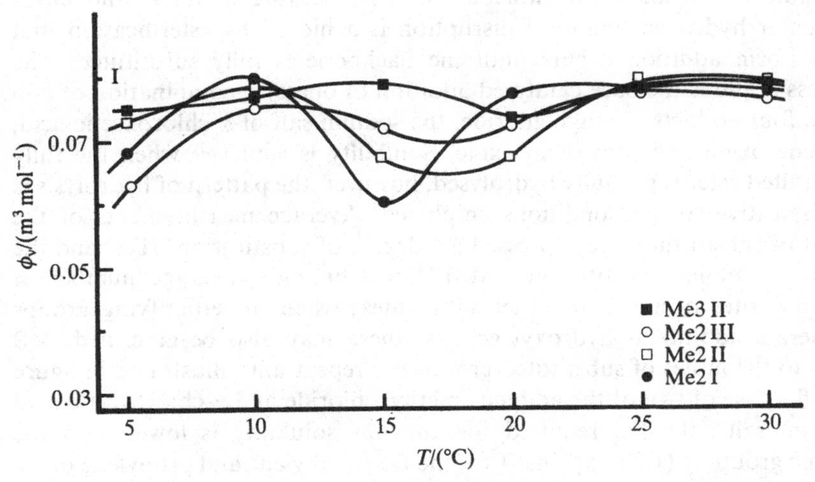

Figure 2. Variations of apparent partial molal volume, ϕ_v with temperature for PVOH/VA copolymers of varying composition (polymer concentration 0.02% w/v) (data from ref. 1).

produce cellulose; if joined via 1–4 axial bonds, α-linkages, then starch is produced, shown in figure 3.

Intricate structural variations, for example the C-2 hydroxyl group being axial rather than equatorial (as in mannose), affect solubility and solution properties, particularly if helical structures result; these characteristics tend

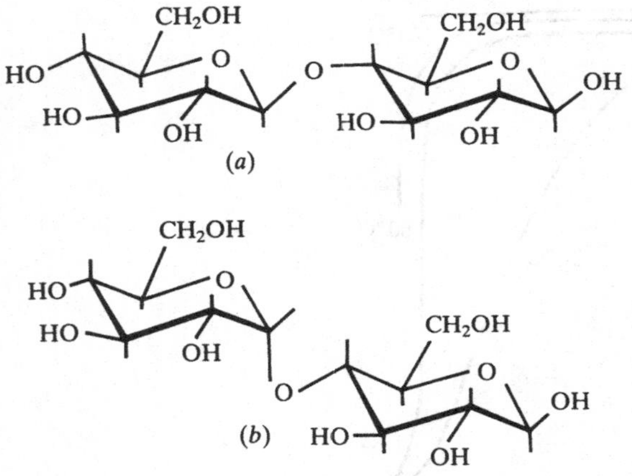

Figure 3. (*a*) 1–4 equatorial, *β*-linkage disaccharide unit. (*b*) 1–4 axial, *α*-linkage disaccharide unit.

to be found in polysaccharides derived from micro-organisms, via fermentation procedures, e.g. Xanthan gum.

Cellulose is insoluble in aqueous solution because of inter- and intramolecular hydrogen bonding; disruption is achieved by esterification, but nonuniform addition occurs until the backbone is fully substituted. The process requires the base catalysed addition of one or a combination of two of the four adducts, methyl chloride, the sodium salt of *α*-chloroacetic acid, ethylene oxide and propylene oxide. Solubility is achieved when the fully substituted ester is partially hydrolysed, however, the pattern of hydrolysis is very sensitive to the conditions employed. Average measurements of the extent of substitution are expressed as 'degree of substitution' (DS) and the extent of 'molar substitution' (MS). DS defines the average number of hydroxyl sites occupied by ether substitutes; when the etherifying groups regenerate additional hydroxyl groups, these may also be etherified. MS refers to the moles of substituted groups per repeat unit, illustrated in figure 4. In the case of two of the adducts, methyl chloride and *α*-chloroacetic acid (sodium salt), the DS required for aqueous solubility is lower with the anionic grouping (1.3 as against 1.6); the DS of ethylene and propylene oxide is difficult to determine because the addition of each adduct leads to the generation of new reaction sites. The MS of adduct per glucopyranosyl unit, however, can be determined experimentally as the amount of oxide derivatives[2].

It is therefore also the case, as with synthetic polymers, that different samples of the same semi-synthetic polymer, whilst having similar values of DS and/or MS may have different solution behaviour, even if their average molar masses are the same.

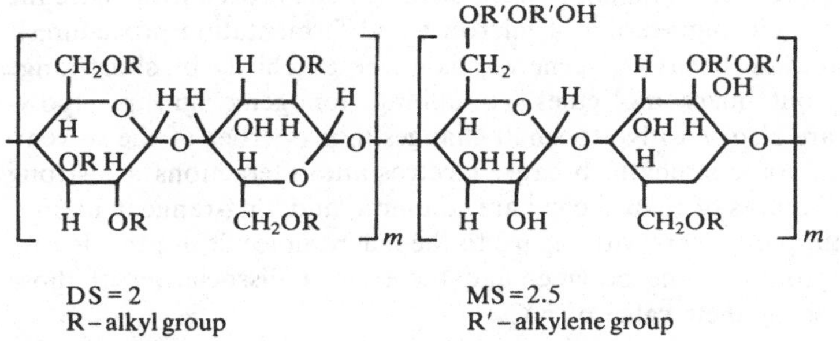

DS = 2
R – alkyl group

MS = 2.5
R′ – alkylene group

Figure 4. The degree of substitution (DS value) and the extent of molar substitution (MS value).

Further similarities in the character of semi-synthetic and synthetic water-soluble polymers occur with chain branching and the presence of ionising groups; branching on the chain tends to disrupt intermolecular associations and hence promote solubility, for example in amylopectin, glycogen and amylose. Polar groups such as $—COO^-$, $—SO_3H^-$, or $—SO_4^{2-}$, being readily solvated, also lead to solubility.

A third structural feature, unavailable to synthetic polymers, is interunit positional bonding; the most common polysaccharides are formed by condensation via the 1–4 link, however, bonding via a 1–3 link produces lower molecular symmetry and therefore better solubility. Linkage via a 1–6 link gives a dramatic increase in aqueous solubility, due to the C-6 carbon atom being external to the pyranosyl ring, thus allowing an entropic contribution to solubility from rotational freedom; a common example of a 1–6 type polymer being the Dextran polysaccharides.

From the discussion so far it is clear that, for both synthetic and semi-synthetic polymers, it is possible for any single sample to contain considerable variations in molar mass, the extent of substitution/copolymerisation and the arrangement of that substituent – it is therefore essential that material for investigation be characterised as completely as possible. Unfortunately much, very interesting, reported work does not satisfy these criteria and the caveat of possible variability must be borne in mind when perusing the literature, including this review!

It is a scientific irony that much of the work reported upon the solution behaviour of water-soluble polymers was undertaken on a basis of 'simplicity' of the macromolecular structure when compared with natural or biopolymers – the results obtained were expected to lead to a greater understanding of these 'more complicated' macromolecules. Although biopolymers contain much greater numbers of different monomers, for example amino acids in proteins, natural processes can, however, replicate the polymer with great exactitude – fortunately for many living systems! This capability is quite outside that of present day synthetic procedures, hence the resorting to genetic engineering of microbes and fermentation procedures!

Polymer solution behaviour generally is governed chiefly by short-range interactions, but macromolecules containing ionogenic groups (polyelectrolytes) are also sensitive to small changes in properties of the solvent, such as pH or ionic strength, because electrostatic interactions are strong (potential differences of 1000 V cm^{-1} are common) and long-range in nature; thus important parameters with regard to the macromolecule are the density of ionisable groups in the polymer and the ease of dissociation of those groups, defined by their value of pK_a.

The solubility of nonionic polymers derives from the presence of polar groups which interact directly, via hydrogen bonds, with the solvent – hydrophilic effects; these effects may, however, be counterbalanced by a tendency of the polymer to inter- and intramolecular hydrogen bonding, leading to crystallisation and insolubility, *viz*, cellulose and PVOH.

An important additional factor in determining the solubility of water-soluble polymers, both ionic and nonionic, particularly when the solubility/temperature relationship is being observed, is hydrophobic interaction/hydration, arising from the unique behaviour of liquid water, rather than 'direct' polymer/solvent interactions.

When a polymer poly(M) dissolves in a solvent S, three forms of interaction are possible, involving the polymer segments M and solvent molecules S, they are M—M, S—S and M—S. The most obvious differences between aqueous and nonaqueous solutions of polymers arise from such interactions. 'Normal' behaviour, of non-aqueous polymer solutions, is an endothermic heat of dilution and precipitation (if it occurs at all, at accessible temperatures) occurs on cooling; some water-soluble polymer solutions (PAA, PAAm, PMAAm) show similar behaviour but in most cases water-soluble polymer solutions exhibit exothermic heats of dilution and precipitate out on heating.

In a few cases the temperature/solubility relationship is even more complex, the polymer solution exhibiting both an upper and a lower critical solution temperature (UCST and LCST respectively); figure 5 illustrates the phase behaviour for a PVOH/VA copolymer, with 7% residual VA, as a function of composition.

For all the solutions, observed behaviour is the result of a balance (often extremely delicate) in the equilibrium

$$\text{M—M} + \text{S—S} \rightleftharpoons 2\,\text{M—S}$$

and only a small shift is required for behaviour to move from 'normal' to 'abnormal' behaviour, for example PAAm to PMAAm, hydrophilic to hydrophobic behaviour. It should also be noted that the M—M contacts referred to are not specific to those of adjacent monomer units but may be between units well separated on the same macromolecule or between units on

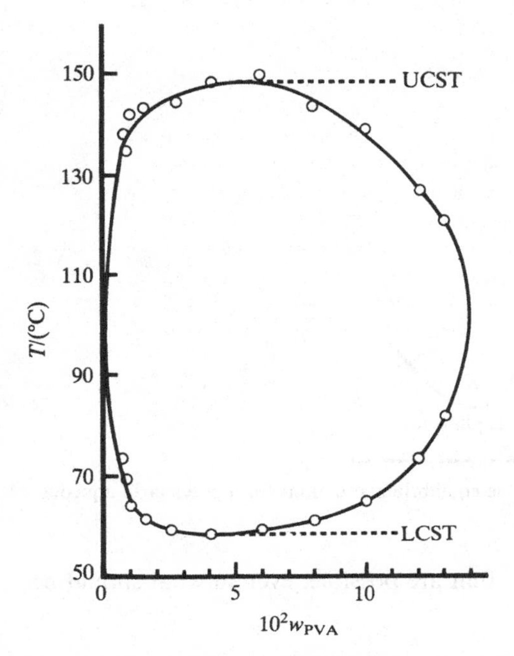

Figure 5. Upper and lower critical solution temperatures of a PVOH/VA copolymer, 7% residual VA content, in aqueous solution ($M = 140000$), as a function of the mass fraction concentration of the polymer (data from ref. 149).

different molecules (intra- or intermolecular), leading possibly, in the first case to a change in conformation, or in the second to aggregation.

When a small molecular cosolute A is also present then two additional interactions become possible, A—S and A—M; if the A—M interaction is greater than A—S and M—S then, even at low cosolute concentrations 'binding' will occur. If no such direct interaction occurs then A—S and M—S interactions can lead to the phenomenon of 'salting out' of the polymer – a quality of solvent effect, a common example of this being the precipitation of proteins by the addition of ammonium sulphate.

When two polymers, poly(M) and poly(N) are present in the solvent S, then the three basic interactions are joined by the further interactions, N—N, M—N and N—S. A modification of this process occurs in the case of polymer adsorption to a substrate, when poly(N) is replaced by the adsorbent B, leading to the interactions M—B and B—S. These last two situations, in their turn, can be complicated by the presence of the small molecule cosolute A.

The broad sub-divisions of interaction will be dealt with in greater detail in subsequent sections, but the brief outline of this introduction serves to show the complex nature of the solution behaviour of water-soluble

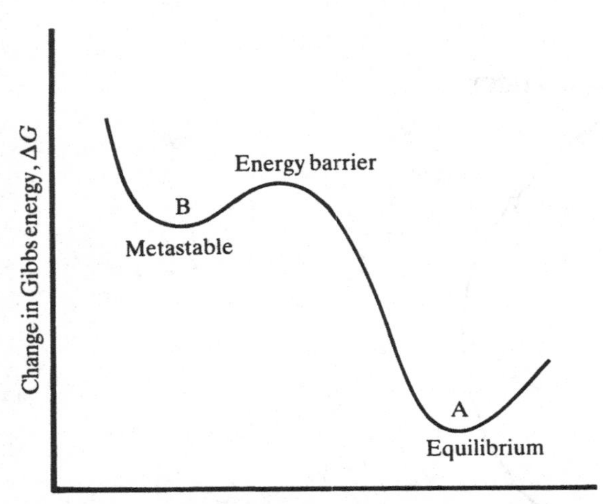

Figure 6. Illustration of true and metastable equilibrium situations for a polymer in aqueous solution (data from ref. 150).

polymers and the many variations that are possible, even in what should be closely similar situations.

2. Thermodynamic behaviour

The thermodynamic functions of state define the equilibrium properties of a solution and should, of course, depend only upon the present state, not its previous history – this is not necessarily the case with water-soluble polymers. In solution behaviour it is assumed that equilibrium is achieved quickly, but in the case of polymer solutions this may not be the case, equilibrium might take hours, weeks, months or years to achieve, in fact it may never achieve the true thermodynamic equilibrium state, depicted by state A in figure 6. State B represents a metastable state and if the energy barrier between the two is very high, interconversion might never be achieved; solutions of 'fully' hydrolysed PVOH are an illustration of this problem – a solution may be obtained under a particular set of experimental conditions, but a change of experimental conditions, again producing a solution, results in a different value for a physical parameter, such as viscosity. Previous treatment of the polymer, such as heating, has a marked effect upon the 'solubility' of fully hydrolysed PVOH.

The physical properties of aqueous polymer solutions are also often found to be time dependent; PVOH (fully hydrolysed) solutions show very marked time dependence, 'ageing', due to structure formation in the solution. Here it has been suggested that a small quantity of n-propanol (2%) in the solution ensures true solution behaviour with no viscosity ageing effects [3], recent work, however [4], has shown by photo-correlation, light-scattering,

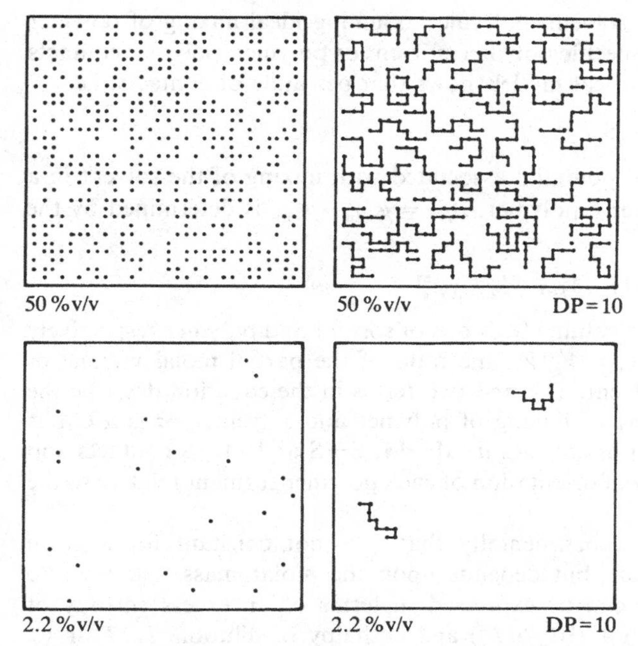

Figure 7. Possible occupation sites for a polymer (DP of 10) on a lattice structure of 900 sites, at concentrations of 50% v/v and 2.2% v/v, compared with similar concentrations of a small molecular solute (data from ref. 150).

viscosity, osmotic pressure and density studies that the polymer is not truly molecularly dispersed.

In all experimental investigations of aqueous solutions of polymers it should always be the case that the solution should be subjected to both heating and cooling prior to investigation; if a constant value of the measured property is obtained when the equilibrium temperature is achieved, then the solution may be at the true equilibrium – or at the very least, a reproducible metastable state.

The fundamental difference that exists between macromolecules and small molecules in solution is illustrated in figure 7, which shows a regular lattice of 900 equivalent sites; a solution of a small molecule, each occupying one site is compared with a small polymer (DP of 10) occupying ten adjacent sites. In a concentrated solution, volume fraction ϕ of 0.5, the polymer solution clearly contains larger units of 'empty space', a difference which becomes even more striking at higher dilutions ($\phi = 0.022$). In the small molecule solution collisions between solute molecules are much more likely than is the case in the polymer solution, where occupation of adjacent sites by polymer segments is likely to involve intramolecular effects.

The Flory–Huggins [5] statistical approach to the solubility of polymers is based upon such a lattice model, each point being occupied by either a polymer segment or a solvent molecule; assuming ideal mixing of polymer and solvent, when no specific polymer/solvent or polymer/polymer contacts are preferred, the statistical model shows that per mole of contacts, i.e.

$$\tfrac{1}{2}\text{M—M} + \tfrac{1}{2}\text{S—S} \rightleftharpoons \text{M—S}$$

the change in chemical potential associated with mixing of the solvent in a solution of a polydisperse polymer, $\Delta\mu_1 = \mu_{\text{soln}} - \mu_{\text{solv}}$ is determined by the equation

$$\Delta\mu_1 = RT[\ln(1-\phi_1) + (1-1/\bar{X}_n)\phi_2 + \chi\phi_2^2]$$

where θ_1 and θ_2 are the volume fractions of solvent and polymer respectively in the solution, and $\bar{X}_n = \bar{V}_2/V_1$, the ratio of the partial molal volume of polymer to that of solvent. The first two terms in the equation describe the configurational entropy of mixing of polymer and solvent, $\chi\phi_2^2$ is a Gibbs energy term arising from changes in M—M, S—S and M—S contacts and changes in the entropy of orientation of each polymer segment relative to the preceding segment.

It has been found experimentally that χ is not constant for a given polymer/solvent system, but depends upon the molar mass, temperature and concentration; χ can be expressed in terms of an excess entropy of dilution, a_x or χ_S $(= \chi + T(\mathrm{d}\chi/\mathrm{d}T))$ and enthalpy of dilution, b_x/T or χ_H $(= -T(\mathrm{d}\chi/\mathrm{d}T))$ where

$$\chi = a_x + b_x/T$$

If the polymer solution separates into two phases upon changing the temperature, then at the critical point for phase separation $\chi = \chi_c$ and the critical temperature T_c is given by

$$T_c = b_x/(\chi_c - a_x)$$

Flory–Huggins theory allows only for endothermic dilution of the polymer solution, i.e. that b_x is positive, but for many aqueous polymer solutions b_x is negative and a positive value of T_c is only allowed if a_x is greater than χ_c. χ_c is dependent upon the molar mass of the dissolved polymer (expressed in terms of the segment number x), in the following way

$$\chi_c = \tfrac{1}{2}x(1 + x^{\frac{1}{2}}) \simeq \tfrac{1}{2} + x^2$$

thus, exceeding the χ_c value produces precipitation of molecules of that particular value of x – the basis of fractionation procedures.

The solution behaviour of the polymer is therefore characterised by the value of χ exhibited:

$\chi = 0$: athermal mixing, only the configurational properties of the polymer are important – entropic in nature.

$\chi = \frac{1}{2}$: the critical value of χ, M—M interactions balance M—S interactions; the polymer chains behave ideally in solution – defined as the

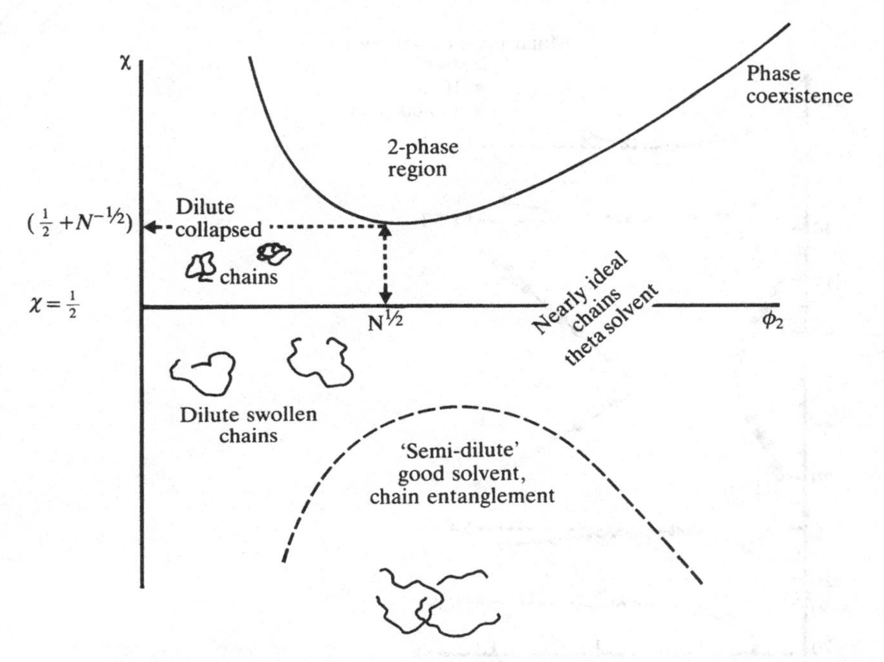

Figure 8. Phase diagram of the Flory–Huggins interaction parameter χ as a function of the concentration ϕ_2 of the polymer, depicting the various solution states of the macromolecule (data from ref. 150).

'theta' condition (a theta solvent and/or a theta temperature). Probably more correctly, this should be described as pseudo-ideal since, although the second virial coefficient (B_2 in the relationship of osmotic pressure to molar mass

$$\pi = cRT\,(1/\bar{M}_1) + B_2c + B_3c^2 + \ldots$$

which usually denotes solute/solute interactions) is zero, B_3 and higher virial coefficients are not necessarily also zero.

$\chi > \frac{1}{2}$: M—M interactions predominate, leading to precipitation – the 'poor solvent' situation.

$0 < \chi < \frac{1}{2}$: M—S interactions preferred, the polymer molecules repel each other sterically in solution, the polymer chains are swollen and the excluded volume effects of the polymer become important – the 'good solvent' condition.

The phase diagram of χ against ϕ_2, figure 8, illustrates the variation of macromolecular behaviour as a function of concentration, the regions of particular interest being:

(i) $\phi_2 < (V_1/\bar{V}_2)^{\frac{1}{2}}$, the dilute solution situation; when $\chi > \frac{1}{2}$, M—M interactions predominate but the concentration is insufficient for

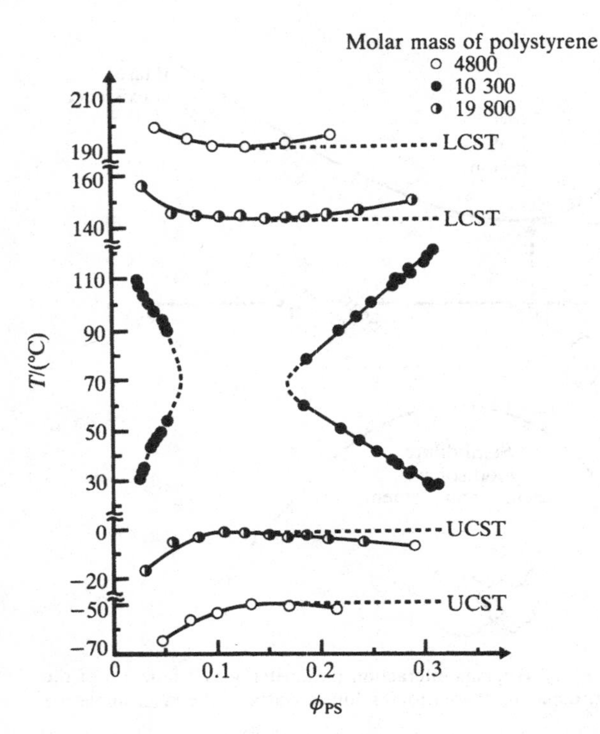

Figure 9. UCSTs and LCSTs for polystyrene in acetone, as a function of volume fraction concentration ϕ_{PS} (data from ref. 149).

precipitation, so the macromolecule exists as contracted coils – intramolecular M—M contacts predominate. If $\chi < \frac{1}{2}$, M—S interactions ensure the polymer is in the form of swollen coils.

(ii) $\phi_2 = (V_1/\bar{V}_2)^{\frac{1}{2}}$ and $\chi = \frac{1}{2} + x^2$: phase separation first occurs – the cloud point.

(iii) $\phi_2 > (V_1/\bar{V}_2)^{\frac{1}{2}}$ with $\chi < \frac{1}{2}$: chain entanglement now occurs between the swollen coils, hence the importance of excluded volume effects in this region.

Variation of ϕ_2 and χ allows investigation of the solution properties of the polymer, either by an alteration of ϕ_2 at a constant value of χ or by altering χ at a constant value of ϕ_2, which may be done by changing the temperature or by modifying the 'quality' of the solvent, from 'good' to 'bad'. The 'normal' behaviour of χ is a decrease in magnitude with increasing temperature, the solvent thus becomes 'better'; on cooling the solvent becomes 'worse' and precipitation occurs – termed θ_- behaviour.

Water-soluble polymers with a high degree of hydrophobic character show a reverse behaviour pattern, the solvent becomes 'worse' on heating and precipitation follows – θ_+ behaviour.

Table 4. *Solubility behaviour of 'normal' and 'abnormal' polymer systems*

Solubility Parameter	Normal	Abnormal
Precipitation occurs on	Cooling	Heating
UCST or LCST	U	L
Thermal nature of mixing and dilution processes	Endothermic	Exothermic
Type of theta temperature	θ_-	θ_+
$d[\eta]dT$	Positive	Negative

If the polymer exhibits both a θ_+ and a θ_- value at accessible temperatures and if $\theta_+ < \theta_-$ then a closed loop is observed, as shown in figure 5; the classical theories of polymer solution behaviour do not allow for such a situation – modified lattice models may account for θ_+ behaviour but only when $\theta_+ > \theta_-$, as shown in figure 9 for a nonaqueous system. The two kinds of solution behaviour, 'normal' and 'abnormal' are summarised briefly in Table 4.

χ values for several typical synthetic water-soluble polymers are shown in figure 10; it is remarkable that values of χ for several of these polymers are very close to the limiting value of 0.5. PMAA for example, which is appreciably hydrophobic in character, is water-soluble by only a very narrow margin; this is, however, in conflict with a reported θ_+ temperature of 56 °C. PVOH on the other hand, with a χ value of approximately 0.48 can have a θ_+ temperature as low as approximately 23 °C. The most likely explanation of the discrepancies is inconsistency in the sample materials – this problem is particularly well illustrated by the data for PVOH shown in figure 5, in this instance both a θ_+ and θ_- value is obtained. These values vary as a function of the polymer concentration; a similar pattern of θ_+ and θ_- can be seen with PEO, where the values of θ_+ and θ_- vary as a function of the molar mass of the polymer – in this case, however, the polymer should be much more soluble since the χ value is of the order of 0.45; at the other extreme the very hydrophilic polymer PAAm($\chi = 0.44$) exhibits only a θ_- value of 6 °C.

The importance of molar mass as a factor influencing cloud point behaviour is shown by the previously mentioned behaviour of PEO. Figure 11, showing the effect of the ratio of VA/VOH diads for several PVOH/VA copolymers of widely varying VA content, between 6 and 24% residual VA, upon the cloud point of the polymer, also illustrates another factor of importance, the influence of the molecular architecture upon the polymer solution behaviour – this should be borne in mind whenever 'copolymer' character may be present in a homopolymer. A further detail of molecular architecture of importance is the tacticity of the polymer, for example atactic PAA and PMAA are water soluble at normal temperatures, whereas the isotactic form tends to precipitate out of solution on standing[6].

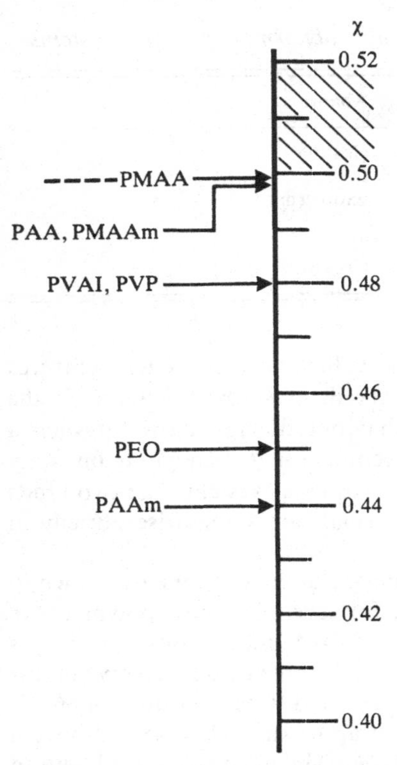

Figure 10. Values of the Flory–Huggins interaction parameter, χ, for dilute solutions of some common water-soluble polymers at 25 °C (data from ref. 151).

The effect of the hydrophilic/hydrophobic balance (HLB) upon the solubility character of the polymer has been studied by Imai and coworkers[7]; increasing the hydrophobic nature of PVOH by partial butyralization consistently lowered the LCST when compared with acetylated[8] and formalated[9] PVOH, at similar degrees of substitution, those for an ethylene substituted PVOH, however, were very similar to the butyralised material. Such behaviour clearly indicates the increasing effect of the more nonpolar substituent group. The root mean square end-to-end distance, $\langle r^2 \rangle^{\frac{1}{2}}$, of the butyralised samples was, however, found to be greater than that of the ethylene substituted material, which according to the authors is due to the increased rigidity of the macromolecular chain, caused by the acetal ring structure and hindrance of intramolecular hydrogen bond formation by the larger butyl side chain. It is also worthy of note that an increasing degree of butyl side chain substitution results in a decrease in the value of the mean square end-to-end distance, indicating a possible intramolecular hydrophobic interaction – detailed information in respect of

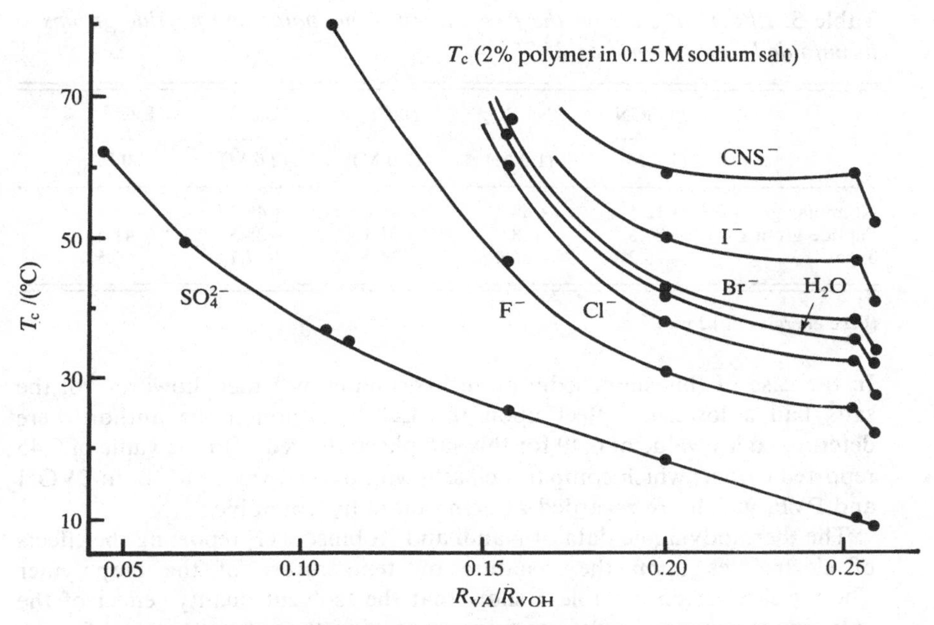

Figure 11. Cloud points of PVOH/VA copolymers of varying composition as a function of VA/VOH diad ratio, R_{VA}/R_{VOH} in water and various salt solutions ($c_s = 0.15$ M) (data from ref. 152).

the distribution of side chain groups within the macromolecule would be most interesting in this context.

The effect of solvent quality upon the solubility of water-soluble polymers is illustrated in figure 11, which shows the consequence of electrolyte addition upon the critical solution temperature of the VA/VOH copolymers as a function of the VA/VOH diads; a clear Lyotropic or Hofmeister pattern of behaviour is observed, with the values for well-known 'salting-out' ions lying below the curve for water (the solvent becoming 'worse'), whilst those for 'salting in' lie above the curve (the solvent becoming 'better'). As in the case of water, the LCST decreases as a function of increasing VA/VOH diad ratio, indicating the increasing importance of hydrophobicity – one curious effect, however, is the discontinuity observed for ratio values greater than 0.25, becoming more obvious with increasing 'salting-in' character of the electrolyte.

Ataman and Boucher[10] have reported the effects of a wide range of salts upon the LCST of a narrow molecular mass fraction of the more water-soluble polymer PEO (20000 ± 700), again a Hofmeister series of effect was observed, viz:

$$PO_4^{3-} > SO_4^{2-} = CO_3^{2-} > S_2O_3^{2-} > F^- > CH_3COO^- > Cl^- > NO_3^-$$

Table 5. *Effects of salts on the free energy of nonpolar and peptide groups in unfolded ribonuclease at 25 °C*

	NaSCN (2.0 M)	Na_2SO_4 (1.0 M)	NaCl (2.0 M)	KCl (2.0 M)	LiCl (2.0 M)
Nonpolar groups	+12.6	+58.4	+46.6	+46.2	+35.0
Peptide groups	−87.8	−8.0	−31.1	−28.5	−41.6
Total	−75.2	+50.4	+15.5	+17.7	−6.5

(Free energies in kJ mol^{-1})

In the case of this supposedly more hydrophilic polymer, however, all the salts had a lowering effect upon the LCST, although the authors here determined a χ value of 0.49 for this sample compared with the value of 0.45 reported earlier, which compares closely with quoted values for both PVOH and PVP, which are regarded as being more hydrophobic.

The thermodynamic data of Nandi and Robinson[11] reporting the effects of electrolytes upon the denaturation temperature of the biopolymer ribonuclease, given in table 5 show that the 'solvent quality' effect of the SO_4^{2-} ion, for example, derives from its making the solvent 'worse' for the nonpolar groups of the macromolecule, whilst the SCN^- ion appears to make the solvent 'better' for the polar groups of the polymer.

The effect of cosolute character upon the cloud point of the water-soluble polymer can become particularly complicated if the cosolute itself has mixed hydrophilic/hydrophobic character; figure 12, as an example illustrates the effect of increasing alkyl group size in a homologous series of tetra-alkylammonium halides upon the LCST of a sample of HPMC. Two effects may be observed; firstly that, as expected, solvent quality improves from chloride to iodide and secondly, it improves with increasing size of the alkyl group, up to *t*-butylammonium halide, before decreasing with the pentyl salt, the decrease being particularly dramatic with the bromide ion and less so with the chloride ion. Similar behaviour is also observed with PVOH/VA copolymers of varying molecular architecture, this effect is therefore clearly caused by the critical size of this cationic species, for reasons at the moment unknown.

PVP is a particularly interesting water-soluble polymer since it is a stable homopolymer, not subject to hydrolysis reactions as is the case with PAAm, and it shows excellent solubility characteristics in water, solution only becoming difficult with increasing concentration due to the considerable rise in viscosity; the χ value of 0.48, however, indicates that water is not a very good solvent for PVP. The conflict probably arises from the hydrophilic behaviour of the cyclic tertiary amide group and hydrophobic hydration involving the methylene groups of the ring structure and the hydrocarbon

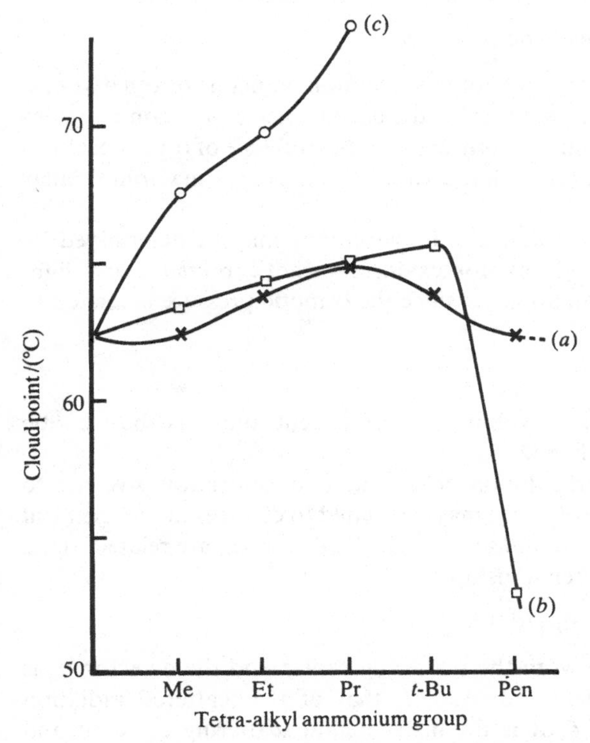

Figure 12. Cloud points of a 1% w/v solution of HPMC in 0.15 M solutions of tetra-alkylammonium halides: (*a*) chloride, (*b*) bromide, (*c*) iodide (data from ref. 153).

backbone. Garvey and Robb[12] have reported heat of dilution (ΔH_d) studies on PVP, from which the enthalpy parameter χ_H may be obtained;

$$\Delta H_d = RT\phi_1\phi_2\Delta n\chi_H$$

ϕ_1 and ϕ_2 are the volume fractions of polymer in solution before and after dilution, Δn is the number of moles of solvent added. The limiting value of $\chi_H^\circ = -0.036$ is probably an indication of interaction of water with the carbonyl group; determination of χ_H° in 0.5 M $(NH_4)_2SO_4$ solution produced a value of -0.11, even more negative than water! Since 0.86 M $(NH_4)_2SO_4$ solution precipitates PVP this indicates a major entropic contribution to the salting out process, due to negative adsorption of sulphate ions by the polymer molecule. This in turn results in an increase of water structure around the non-polar regions of the polymer – a result in agreement with the thermodynamic evidence of Nandi and Robinson[11].

3. Solution behaviour of nonionic polymers

The fundamental property of any polymer solution, aqueous or otherwise, is the chemical potential of the solvent μ_1; the dependence of μ_1 upon variables such as solute concentration, temperature and the presence of other cosolutes can yield valuable information with regard to the state of the macromolecular species in solution.

The solvent chemical potential and its variations may be determined by several experimental procedures, for example osmotic pressure and light scattering yield information about μ_1 since the osmotic pressure is related to μ_1 by the relationship

$$\pi \bar{V}_1 = -\Delta\mu_1$$

where $\bar{V}_1$ is the partial molal volume of the solvent and π is the osmotic pressure, as previously defined.

Local microvariations in solution density and concentration give rise to fluctuations in local values of π and may be related to changes in the chemical potential of the solvent (for dilute solutions) which, in turn, are related to the scattering of light via the relationship

$$I_\theta r^2/I_0 = R_\theta = -Kc\bar{V}_1 RT/(\mathrm{d}\mu_1/\mathrm{d}c)$$

where r is the distance between the scattering centre and the detector, I_0 is the intensity of the incident radiation, I_θ that of the scattered radiation at the scattering angle of θ, K is the instrumental scattering constant and c is the concentration.

Scattered light measurements as a function of polymer concentration and scattering angle allow the construction of a Zimm diagram, where Kc/R_θ is plotted against $\sin^2(\theta/2)+$ constant. Extrapolation to infinite dilution and zero scattering angle provides information on the mass average molar mass since

$$Kc/R_\theta(1+\cos\theta) = (1/\bar{M}_\mathrm{w})P_\theta^{-1} + B_2 c + B_3 c^2 + \ldots$$

Information regarding particle shape may also be obtained from the angular dependence of the particle scattering factor P where

$$P_\theta^{-1} = 1 + 16\pi^2/3\lambda^2\langle\bar{s}\rangle^2 \sin^2(\theta/2)$$

and $\langle\bar{s}\rangle^2$ is the mean square radius of gyration; the magnitude of the second virial coefficient, obtained from the initial gradient of the zero angle line yields information as to the quality of the solvent, positive values indicating a good solvent. The size of the macromolecule in solution, given by $\langle\bar{s}\rangle^2$, is obtained from the intercept and initial slope of the zero concentration line.

Comparison of the number average molar mass $\bar{M}_\mathrm{n}$ from osmotic pressure measurements with the mass average molar mass, from light scattering data

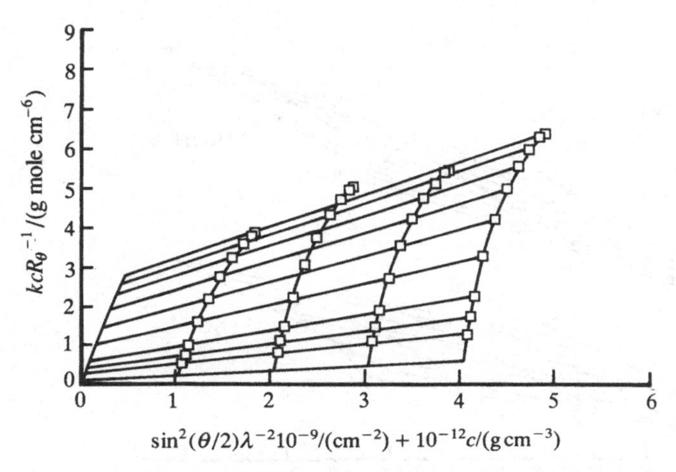

Figure 13. Zimm plot of the system PVOH/water/2% w/w n-propanol, $T = 25$ °C (data from ref. 4).

gives M_n/M_w, the polydispersity ratio – a very useful indication of the spread of molecular mass within the sample under investigation.

Figure 13 illustrates a Zimm plot of light-scattering data obtained using a sample of 'fully' hydrolysed PVOH[1]; curvature of a Zimm plot is usually an indication of aggregation[13] in the polymer solution, the value for $\overline{M}_w$ obtained in this case being 200000, which was markedly different from the value of 100000, expected from studies of the parent polymer PVA in methanol solution. Difficulties were also experienced in obtaining osmotic pressure measurements upon this material – the molecular mass in this case being approximately 25000, but measurements upon the parent PVA in methanol gave a value of 90000. The data therefore clearly indicate a problem with aggregation in the solution; Mrckvickova, Prokopova and Quadrat[14] from classical light-scattering studies have further shown that the ageing of fully hydrolysed PVOH results in the growth of supramolar aggregates. Aggregation within PVOH solutions is not, however, restricted to the fully hydrolysed material, with its well-known solubility problems. Recent light-scattering data[15] obtained using a PVOH/VA copolymer containing 20% residual VA units, revealed the existence of aggregates in the solution, which formed over a narrow range of concentration – effectively a 'critical micelle concentration' of the polymer. Light-scattering and high precision density measurements on very dilute solutions of this polymer[16] have also shown that the macromolecule undergoes a marked change in conformation as a function of temperature, the change occurring at around 15 °C; analysis of the partial molal expansibility values showed that the temperature dependence could be divided into three regions, below 10 °C, between 10 and 15 °C and above 15 °C. Below 10 °C the major influence

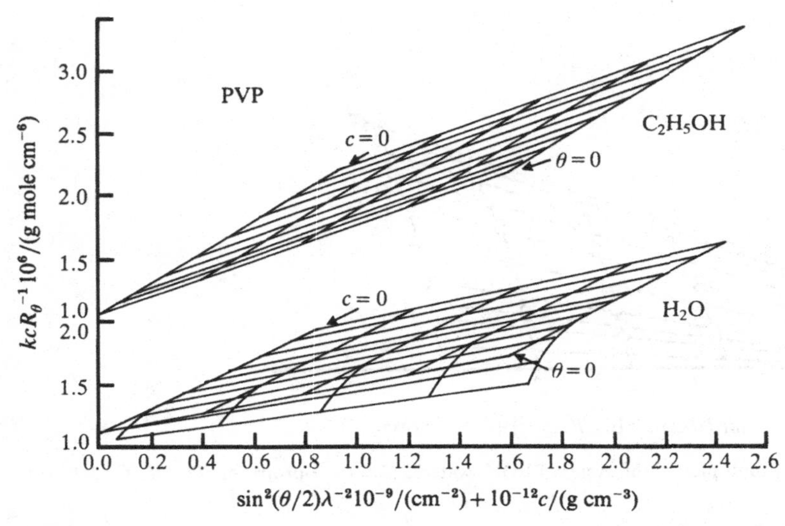

Figure 14. Zimm plot for PVP in water and ethanol solution, $T = 25$ °C (data from ref. 150).

affecting solution behaviour was hydrophilic in nature, increasing temperature resulted in an increasing hydrophobic contribution, which achieved a maximum at 15 °C; above 15 °C hydrophilic effects again become of importance. It is particularly interesting in this respect to consider the low temperature heat capacity studies of Franks and Wakabayshi[17] on PVP, which show that, with decreasing temperature, from 10 to −40 °C, hydrophilic effects become increasingly dominant.

Studies of the effects of low concentrations of electrolytes upon the conformation change of the PVOH/VA copolymer[16] showed that the presence of the SO_4^{2-} ion has the effect of shifting the temperature of the conformational change to lower values, perhaps as a result of an increased hydrophobic effect, as outlined earlier; the presence of lithium ion (a salting-in agent) has the effect of moving the temperature of the change upwards. Aggregation behaviour is also likely in solutions of PVP; figure 14 illustrates Zimm plots for PVP in ethanol and water, the data for ethanol solutions show classical behaviour but those for water again show distortion, characteristic of aggregation, a conclusion supported by the data of Garvey and Robb[12]. It should be noted that in all the cases considered the value of the second virial coefficient is positive, water is therefore a satisfactory solvent – this, however, does not prevent aggregation, even at very low polymer concentrations.

Light scattering experiments on a series of PAAms of varying molar mass, reported in a comprehensive review by Kulicke, Kniewski and Klein[18] show that for $\overline{M}_w$ values of 8×10^5 and below Zimm plots are of the expected regular form but for 4×10^6 and greater, curvature is observed; the second

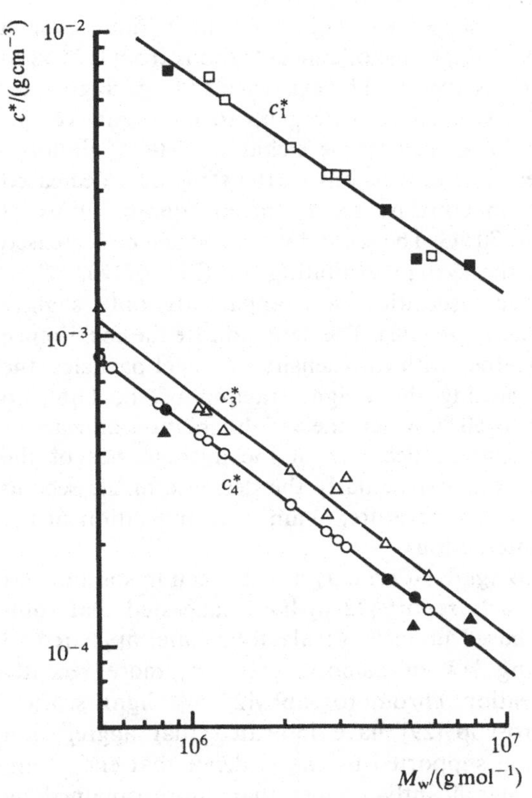

Figure 15. Critical aggregation concentrations of PAAm in water at 25 °C, dependence upon molar mass; c_1^* from viscometry data, c_3^* from light scattering data, c_4^* for 'good' solvents (data from ref. 18).

virial coefficient also decreases, these results again indicating the onset of an aggregation process. Figure 15 illustrates the critical concentration for onset of aggregation, determined from viscometry and light scattering, as a function of molar mass; the differences between c_1^* and c_3^* arising from the different techniques, whilst c_4^* is derived from the reptition concept[19] for 'good' solvents. It follows from this work that any determination of the molar mass must be conducted at concentrations below c_1^*.

Ageing effects have also been observed in aqueous solutions of PAAm, leading to time dependent viscosity changes, but without change of molar mass; Kulicke *et al* suggest that the ageing phenomenon is a consequence of a conformational change, controlled via intramolecular hydrogen bonds, leading to a more compact flexible coil structure.

The possibility of aggregation in dilute polymer solutions appears, from the data discussed, to be a widespread phenomenon and the light-scattering data on PEO solutions reported by Polik and Burchard[20] lends further support to this possibility; these workers obtained their data over a temperature range of 20–90 °C, for a range of concentrations from 0.25% to 2.0% for a sample of molar mass 2×10^4. The existence of large aggregates was noted, which increased in size upon heating up to 60 °C, above this temperature M_w decreased; at 40 °C the data were characteristic of globular, fairly monodisperse structures. The dimensions of the structures decreased with increasing temperature, in contrast to the mean square radius of gyration, which increased up to 70 °C. The second virial coefficient decreased with increasing temperature, the system exhibiting a LCST of 102 °C; it should be noted that intrinsic viscosities were apparently only slightly influenced by the presence of the aggregates. The data indicate the coexistence of high-density spherulites together with low-density microgel particles, the former melting on heating, causing the weight fraction of the latter to increase with temperature up to 60 °C when the weight fraction appears to decrease, but without any apparent decrease in the average size of the microgel particles. Such behaviour, particularly the decrease in the second virial coefficient with increasing temperature, again is an indication of the importance of hydrophobic interactions.

Similar complications due to aggregation may also be seen in solutions of semi-synthetic polymers[21]; early reports[22–4] have suggested that solutions of pectin, a copolymer based upon (1–4)galactouronate, appeared to follow the van't Hoff limiting law of osmotic pressure; more recently measurements by gel permeation chromotography[25,26], light scattering[27,28] and electron microscopy[29] have indicated that aggregation occurs in solution, a conclusion supported by the evidence that end-group titration gave values for $\bar{M}_n$ significantly smaller than those obtained by membrane osmometry[30].

Measurements by Fishman, Pepper and Pfeffer[21] on a series of methyl esterified pectins also revealed differences between $\bar{M}_n$ determined by membrane osmometry and end-group titration; following heat treatment, however, and a time lapse of several days, a concentration dependent aggregation process was discerned, illustrated in figure 16 – the values of π/c at the intercept being calculated from $\bar{M}_n$ values obtained by end-group titration; interestingly, the authors suggest that the systems form non-ideal aggregates under the influence of large free-energy changes.

Fluorescence spectroscopy on pyrene-labelled HPC[31] provides further evidence that semi-synthetic water-soluble polymers exhibit a tendency to association, even at very low concentrations; this tendency undergoes a marked change when the solvent is 'worsened' by the addition of methanol, shown in figure 17. HPC is a relatively hydrophobic polymer, thus the action of the methanol is likely to be disruption of the hydrophobic interactions of the polymer, allowing hydrogen bonding requirements to be fulfilled, leading in turn to a change of conformation. In a subsequent paper Winnick[32] utilised the same technique to investigate the temperature dependence of the

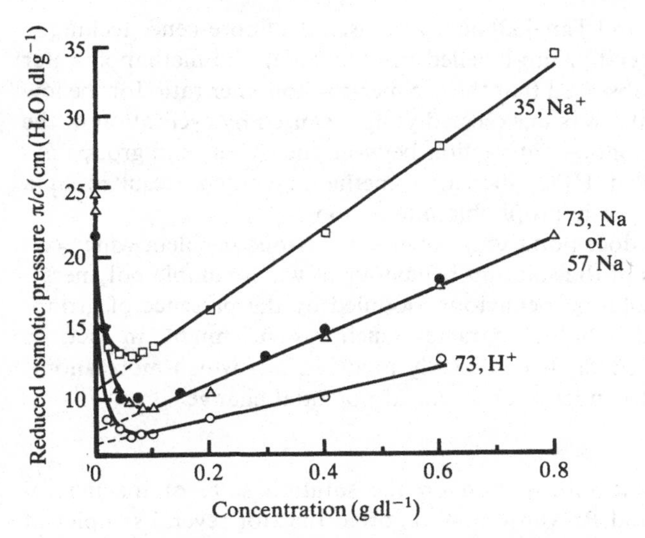

Figure 16. Van t'Hoff plot demonstrating concentration-dependent dissociation of pectin; the values at the intercept are calculated from end-group titration (data from ref. 21).

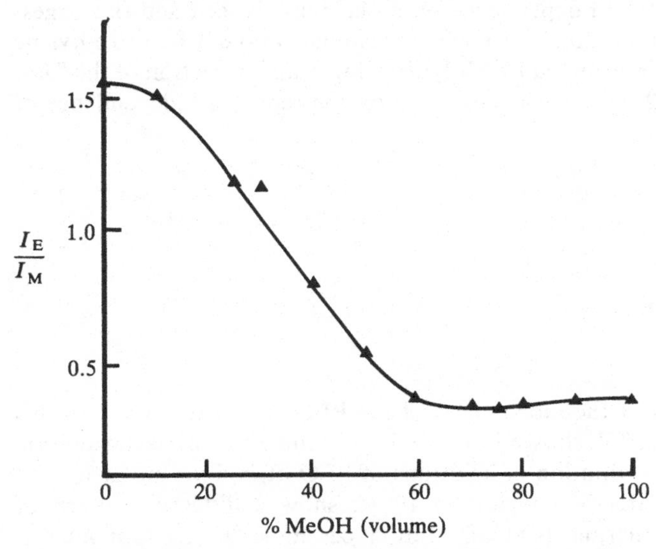

Figure 17. Plot of the excimer/monomer fluorescence intensity ratio I_E/I_M for pyrene-labelled HPC as a function of water/methanol composition (v/v) (data from ref. 31).

fluorescence of pyrene-labelled HPC; the intense excimer emission decreased sharply prior to the cloud point of the modified polymer, confirming the existence of hydrophobic interaction between the pyrene groups on the polymer.

Char, Frank, Gast and Tana[33] have also used the fluorescence technique to investigate pyrene end-group-labelled PEO in water and methanol/water mixtures; this group observed that the excimer-to-monomer ratio for the low molecular weight (4800) was unexpectedly high, caused by cyclisation of the polymer due to hydrophobic interaction between the pyrene end groups. As in the case of labelled HPC, increased methanol content resulted in a decrease of the extent of hydrophobic interaction.

The data reviewed does point to a potentially serious problem which can arise in examinations of the solution behaviour of water-soluble polymers – how far is the true solution behaviour modified by the presence of groups with appreciable hydrophobic character; such systems might, in fact, be more accurately termed 'hydrophobically modified', of which more anon.

The temperature dependence of chemical potential change

$$d[(\Delta\mu_1)/T] = \Delta\bar{H}_1$$

can also yield valuable information on the solution state of the macromolecule; Malcolm and Rowlinson[34] reported that for several samples of differing molar masses of PEO, the heat of dilution (at 80.3 °C) was independent of molar mass, up to solution concentrations of 50%. At higher concentrations of PEO a dependence on molar mass is seen and the largest value for the heat of dilution is seen when the molar ratio of water to ethylene oxide units is 2:1 – interpreted by Molyneux[35] as an interaction of the form shown in diagram 2. In the water-rich region the reported independence of

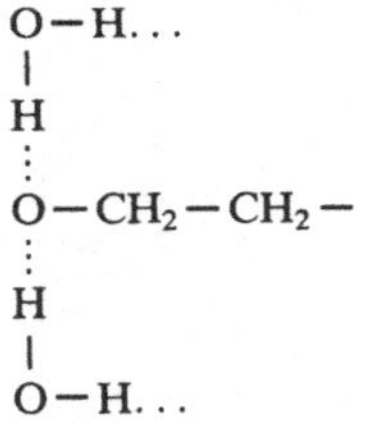

```
O−H...
|
H
⋮
O−CH2−CH2−
⋮
H
|
O−H...
```

Diagram 2

the heat of dilution of the molar mass of the PEO is said to be due to this process. Recent data[36], shown in figure 18, obtained by microcalorimetric studies of the heat of dilution of solutions of PEO of various molar masses in the extremely water-rich region at 10 °C, show a different pattern of behaviour; the exothermic heat of dilution per mole of EO monomer is greatest for the smallest molar mass (200), decreasing with increasing DP, but increasing with increasing concentration; there is therefore a definite dependence on molar mass within the water-rich region, which is apparently reaching a limiting value above a molar mass of 6000, indicating the relative importance of end-group interactions with decreasing size of the oligomer. This conclusion is in agreement with the data of Kagamoto, Murakami and Fujishiro[37], who showed from heats of dilution measurements that the

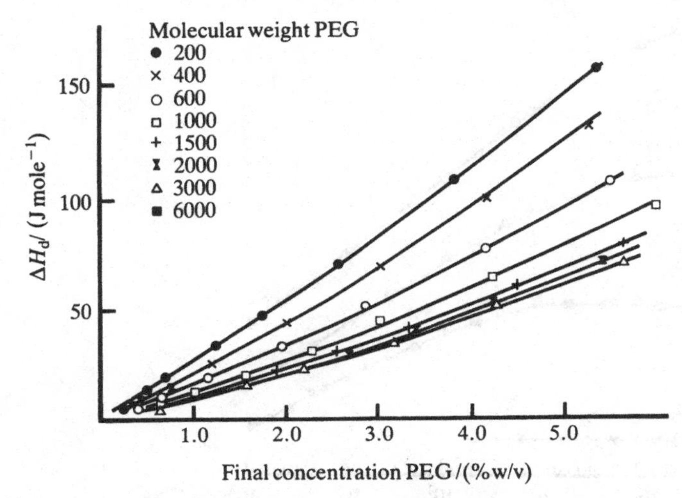

Figure 18. Heat of dilution of PEO as a function of molar mass and polymer concentration at 10 °C; (data from ref. 36).

enthalpy term in the interaction parameter χ_H was most negative (-0.90) for a molar mass of 200, increasing with molar mass to a limiting value of -0.05 for molar masses in excess of 2×10^4.

Reference has been made previously to the phenomenon of 'ageing' in aqueous polymer solutions, particularly with regard to viscosity. Due, most probably, to the historical reason that many of the uses of water-soluble polymers in the industrial context involve their capacity to act as thickeners, controlling the flow properties of the systems, much of the investigation of solution behaviour has centred upon the dynamic properties of the polymer in solution. Intrinsic viscosity values are a well-used parameter for determination of molar mass, via such treatments as the Mark–Houwink equation (5), of the form

$$[\eta] = KM^a$$

where K and a are constants specific to the polymer and the solvent; examination of the literature reveals that frequently discrepancies exist in the values of K and a for all water-soluble polymers – this has been well summarised for PAAm by Kulicke and coworkers[18], values of K ranging from 6.31×10^{-3}–1.0×10^{-2} being reported, whilst 'a' varies from 0.66 to 0.80. Such variations may be caused by polydispersity in the molar mass, crosslinking or chain branching, but the factor that should always be considered in the viscosity of water-soluble polymers is the shear-rate dependence of the solution. A viscosity dependence upon shear-rate may be observable, even in dilute solutions, illustrated clearly in figure 19, and it should be noted that capillary viscometers generate shear-rates of several

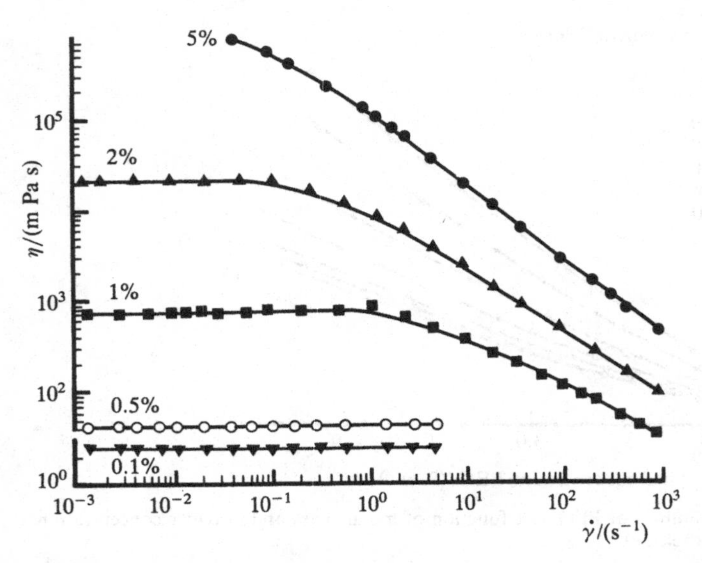

Figure 19. Viscosity as a function of shear-rate for aqueous solutions of PAAm of different concentrations, molar mass 5.3×10^6 g mol^{-1} $T = 25$ °C (data from ref. 18).

hundred reciprocal seconds – care must therefore be exercised in determining the intrinsic viscosity of high molar mass polymer samples. Similar precautions pertain with regard to the Huggins equation (5)

$$\eta_{sp}/c = [\eta] + k_H[\eta]c^2$$

where the estimation of solvent quality is derived from the magnitude of the Huggins coefficient k_H, 0.3 for rod-like molecules in a 'good' solvent, up to about 0.7 under theta conditions, larger values indicating the likelihood of polymer association – here again, however, viscosity ageing phenomena can affect the accuracy of the coefficient, figure 20 illustrates the effect for a sample of PAAm, $M = 5.05 \times 10^6$.

Non-Newtonian shear dependent flow behaviour is not restricted solely to synthetic polymers such as PAAm and PVOH[38], it is a particularly important aspect of the behaviour of what have become known as 'hydrophobically modified polymers'; these materials may have originally been either completely synthetic or semi-synthetic, but which have subsequently been further modified by the attachment of a small percentage of hydrophobic groups (usually less than 1% of the polymer by weight).

Figure 21, from the data of Sau[39], illustrates the influence of hydrophobe content upon the shear-rate dependence of the viscosity of a C_{16}-HEC ($CH_3(CH_2)_{15}$ chain attached by the ethylene oxide group of the polymer). The data of Bock, Siano, Valint and Pace[40] show that the presence of a hydrophobe is the determining factor in the quality of the solvent for that

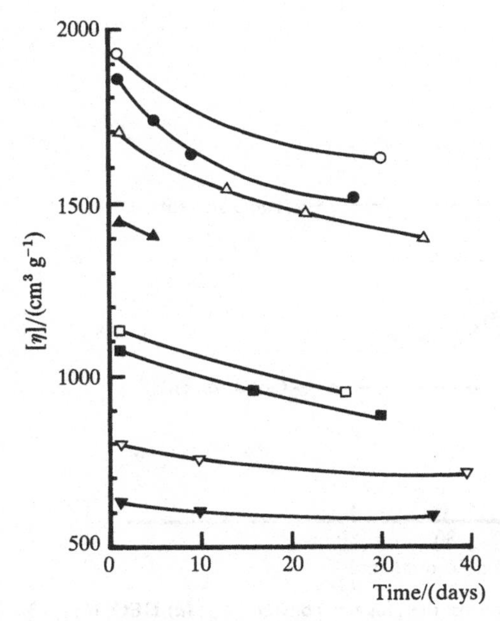

Figure 20. Viscosity ageing behaviour of PAAm – time dependence of the intrinsic viscosity of PAAm in water at 25 °C for a wide variety of molar masses (data from ref. 18).

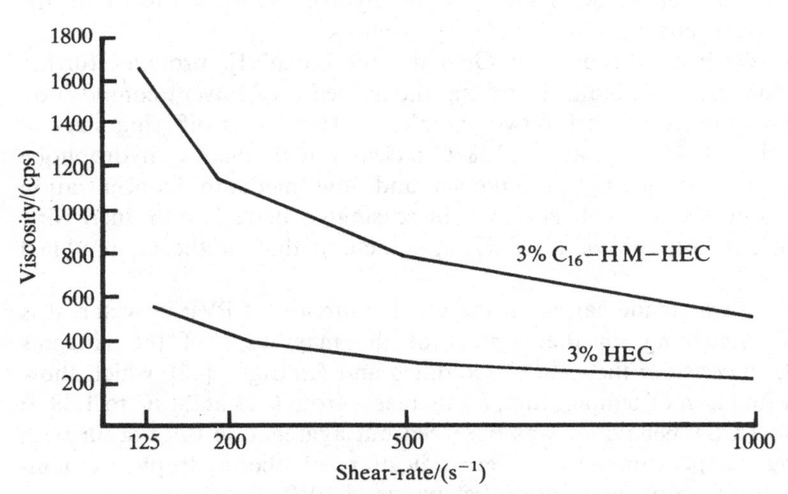

Figure 21. Dependence of viscosity upon applied shear-rate for solutions of HEC and hydrophobically modified HEC (data from ref. 39).

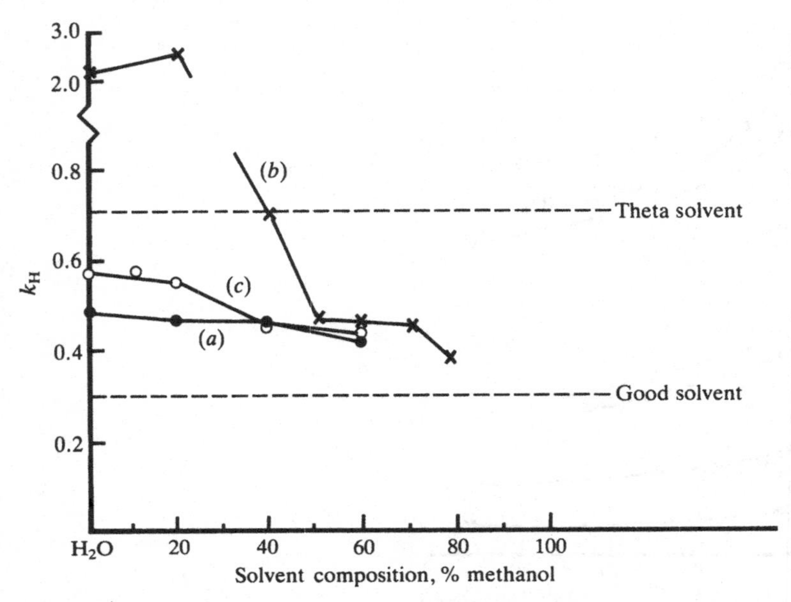

Figure 22. Effect of methanol concentration on the Huggins constant, k_H; (*a*) HEC; (*b*) 0.9% C_{16}HM-HEC; (*c*) 1.0% C_8HM-HEC (data from ref. 41).

polymer, k_H changing from 0.5 for partially hydrolysed PAAm to 1.8 for the polymer containing 1.25% of a C_{16} hydrophobe.

Figure 22, from the data of Gelman and Barth[41], provides further support for this viewpoint, illustrating the influence of solvent composition on the Huggins constant for two samples of HEC with differing sizes of hydrophobes, 0.9% C_{16} and 1.0% C_8; clearly with the C_{16} hydrophobe aggregation is occurring in aqueous and low methanol concentration solvents, but the solvent becomes increasingly 'better' with increasing methanol content, the value of k_H approaching that of the C_8 modified HEC.

Reference was made earlier to the good solubility of PVP in water, it is therefore interesting, in the context of the magnitude of the Huggins constant, to consider the data of Goldfarb and Rodriguez[42], which show that as a function of temperature, k_H increases from 0.23 at 20 °C to 1.48 at 60 °C; no LCST behaviour was observed but aggregation does occur with increasing temperature – a further indication of the hydrophobic contribution to the aqueous solution behaviour of PVP.

The dependence of the viscosity of the polymer solution upon applied shear-rate means that a critical viscosity value pertains which is shifted to lower values of shear-rate with increasing mass and/or concentration; such behaviour is said to be typical of polymer solutions in which the polymer is dissolved in statistical coils, hence a plot of zero shear viscosity against molar

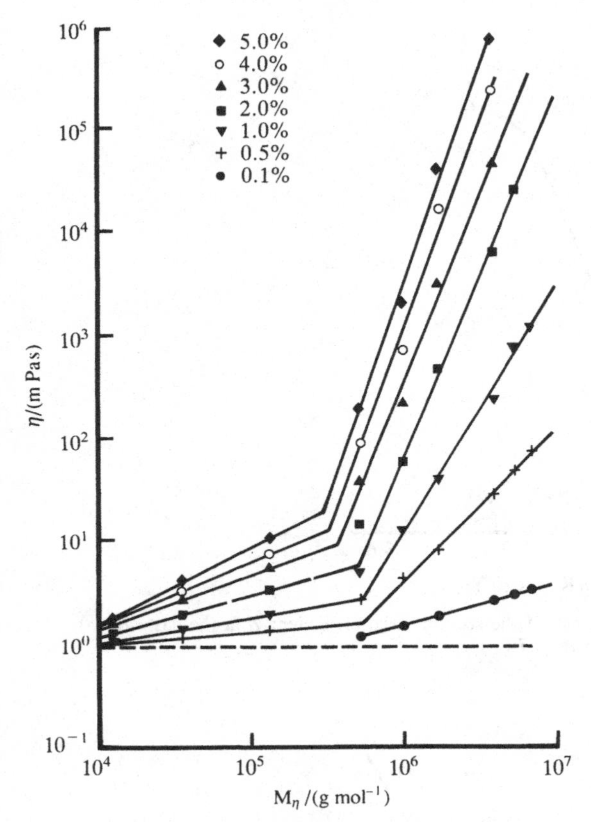

Figure 23. Zero shear viscosity versus molar mass for different concentrations of PAAm in water, $T = 25$ °C (data from ref. 18).

mass should show a typical plot of two straight lines of different slopes, the intersection representing the critical concentration for entanglement, producing a continuous network[43]. Figure 23 illustrates the situation for PAAm of varying molar mass, M; the data also indicate a critical value of molar mass, M_c, 4.4×10^5 for chain entanglement, however, the slopes deviate considerably from those expected on the basis of statistical coil behaviour. 1.0 for $M < M_c$ and 3.4 for $M > M_c$, suggesting that this behaviour may be better explained on the basis of a suspension model. Similar behaviour has been observed with PVOH solutions[44] and treatment of the flow of aqueous solutions of PVOH/VA and PVP on the basis of a flow equation for high volume fraction polymer dispersions[45]

$$\eta/\eta_0 = 1 + 2.5\phi + 10.05\phi^2$$

by means of the substitution $\phi = cV_e$ (η_0 is the viscosity of the solvent, η is

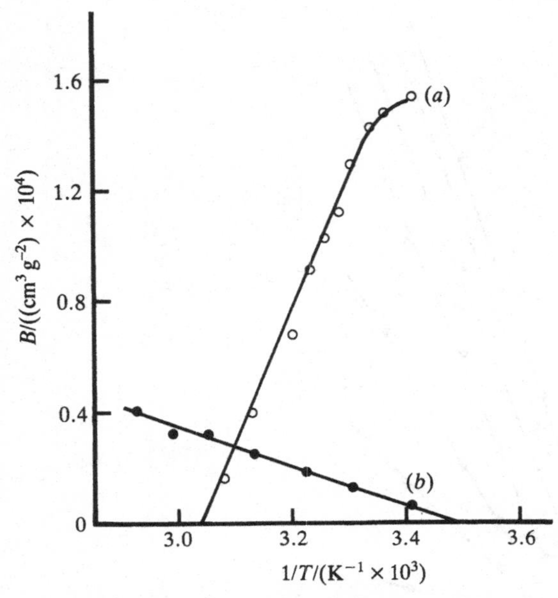

Figure 24. Temperature dependence of the second virial coefficient B in the cases of (*a*) PMAA and (*b*) PAA (data from ref. 47).

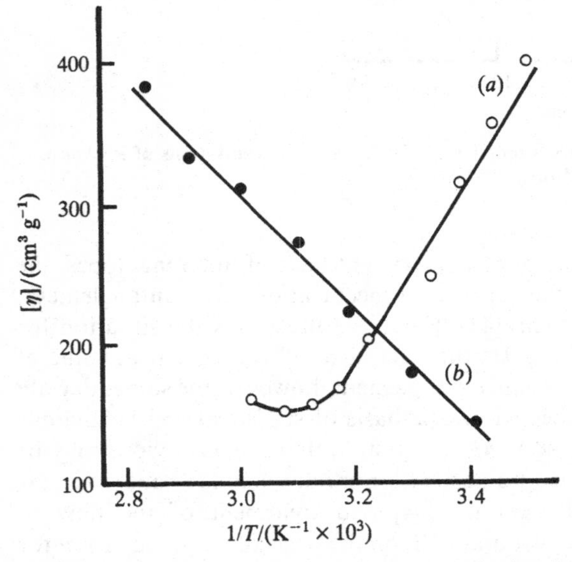

Figure 25. Temperature dependence of the intrinsic viscosity $[\eta]$ in the cases of (*a*) PMAA and (*b*) PAA (data from ref. 47).

the viscosity of the solution, c is the concentration of the polymer in solution and V_e its effective flowing volume) yields results in good agreement with those obtained from adsorption studies[46].

A discussion of the solubility of water-soluble polymers would not be complete without reference to the definitive paper of Silberberg, Eliassaf and Katchalsky[47] relating light-scattering and intrinsic viscosity studies of the related family of acrylic polymers, PAA and PMAA, together with PAAm and PMAAm; the two acids were dissolved in 0.02 M HCl to ensure suppression of the acid dissociation constant and hence avoidance of polyelectrolyte character. The extended studies over a range of concentrations and temperatures resulted in values for the second virial coefficient as a function of temperature, shown in figure 24 and intrinsic viscosity as a function of temperature, shown in figure 25; PMAA is seen to have an LCST of 329 K whilst PAA exhibits a UCST with a value of 287 K – the increased hydrophobicity of the methyl group showing clearly in this context. This situation is also clearly observed in the intrinsic viscosity behaviour of PMAA when compared to PAA. A conclusion drawn from the temperature dependence of the second virial coefficient is that the 'dilution' contribution to the heat of mixing of polymer and water is exothermic in the case of PMAA but endothermic in each of the other three cases; a recent microcalorimetric determination of the heat of dilution of PMAA in the water-rich region[36] however, showed the value to be negligibly small, at least at 283 K, with, admittedly, a much smaller mass, 14000, compared to 2.5×10^5 in this report.

4. Aqueous solution behaviour of polyelectrolytes

A polymer molecule having charged groups attached by covalent linkages to the polymer backbone is termed a 'polyelectrolyte'; if the charged groups are negative, then it is known as a polyanion, positive charge results in a polycation and the presence of both kinds of charged groups gives rise to a polyampholyte. Except in the unique case of the polyampholyte containing equal numbers of positive and negative charges the macromolecule will maintain a residual charge, which is only partially neutralised by the presence of 'bound' or 'condensed' counter or gegenions. The situation is summarised in figure 26, showing that although the polyelectrolyte solution must be electrically neutral overall, this is not necessarily, or even usually, the case in the micro-environment of the polymer backbone.

Examples of the more common synthetic polyelectrolytes, both anionic and cationic are shown in table 1; typical examples of semi-synthetic polyelectrolytes include the sodium salts of CMC, CMHEC, pectic acids and Xanthan gum.

The polyelectrolyte character of many of the examples cited depends upon the degree of neutralisation or protonation of the parent polyacid or base –

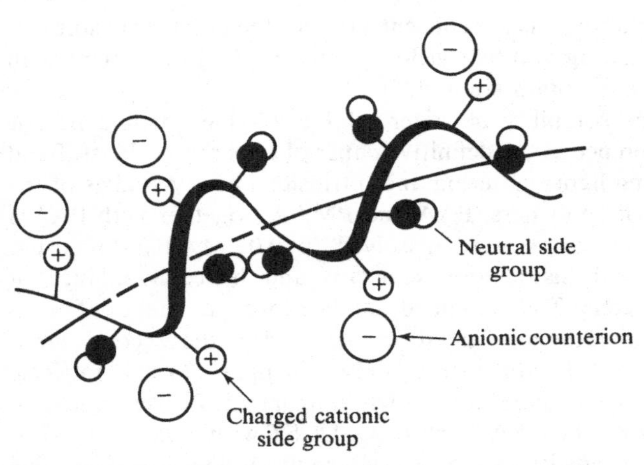

Figure 26. Descriptive picture of a polyion in aqueous solution.

the extent of polyelectrolyte behaviour therefore ranges from essentially nonionic character in the case of the undissociated acid to highly charged behaviour in the case of the fully neutralised salt.

The presence of charge centres along the macromolecular backbone must mean that the whole of the polymer molecule is surrounded by a double layer of electric charge, identical to that surrounding all colloidal particles; however, unlike a normal colloid particle, the density of those charges along the backbone has important consequences for the flexibility of the macromolecule. Figure 27 illustrates qualitatively the two extremes of high and low charge densities, typified by a polyacid with low and high degrees of neutralisation – in the first case a flexible molecule will result, but in the second a rigid rod-like entity will be produced.

The extent of the double layer of charge surrounding the molecule, as with colloidal systems in general, will be sensitive to the concentration of counterions in the surrounding solution; such complex behaviour leads to difficulties in the interpretation of the aqueous solution behaviour of polyelectrolytes. The macromolecular and double layer aspects of behaviour are nonadditive and characterisation of the macromolecule by, for example light-scattering measurements, must take this into account – usually by addition of sufficient electrolyte to suppress double layer behaviour; an electrolyte concentration of about 0.15 M is significant in this respect, below this concentration double layer effects are important whilst higher salt concentrations influence the quality of the solvent. These factors are also, of course, in addition to those observed with nonionic polymers, such as hydrogen bonding and hydrophobic hydration effects.

The effect of charge density and counterion concentration on the dimensions of the polyion can be seen qualitatively in the variation of

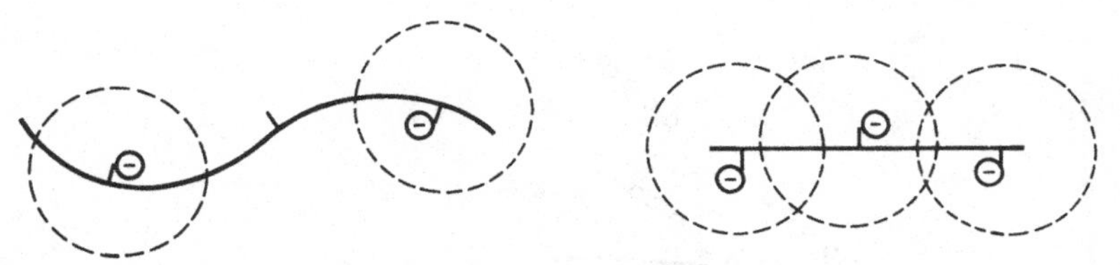

Figure 27. Illustrative example of the flexibility associated with the degree of dissociation of a polyacid.

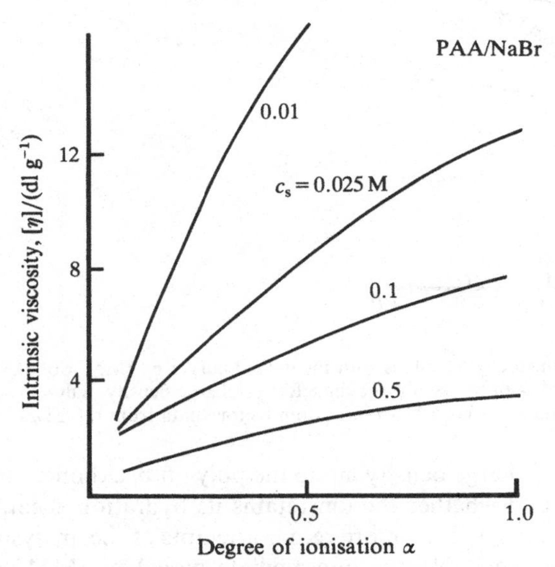

Figure 28. Intrinsic viscosity of PAA in aqueous solution as a function of the degree of ionisation α and increasing sodium bromide concentration (data from ref. 150).

intrinsic viscosity with degree of ionisation and added salt concentration, shown in figure 28, illustrating the neutralisation of PAA in various concentrations of sodium bromide; the rapid decrease in $[\eta]$ with increasing sodium bromide concentration, up to 0.1 M, is associated with the suppression of the charge atmosphere – the molecule becoming less rod-like and hence more flexible, leading to a reduction in $[\eta]$; the much smaller changes in $[\eta]$ at higher salt concentrations are a reflection of the quality of the solvent.

Theoretical models for the state of the polyion and associated counterions have been published by Manning[48], Yoshida[49], and Satoh, Kowashima, Komiyama and Iijima[50]; analyses can be broadly divided on the basis of whether counterions are regarded as 'site-bound' by charge centres or more generally 'condensed' within the close environs of the macromolecule and regarded as being intimately associated with the polyion. Variations on the

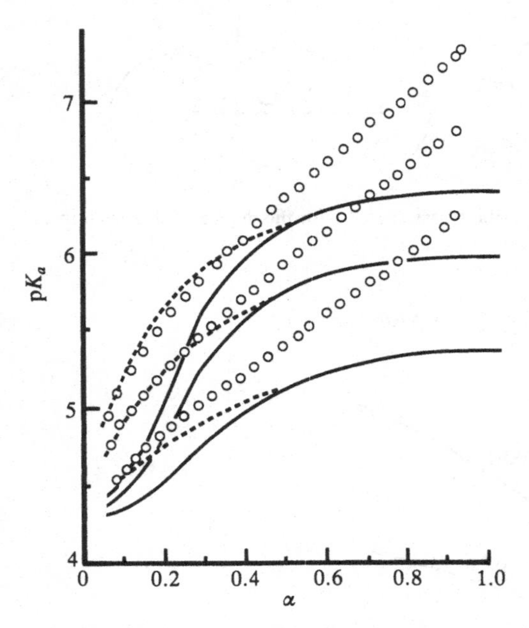

Figure 29. Comparison of experimental pK_a values with theoretical curves by IMM. Solid curves IMM, dashed curves IMM with correction for the effective charge density. Salt concentration values are 0.1 M, 0.02 M, 0.005 M from bottom to top (data from ref. 51).

models include the extent of charge density along the polyion backbone, the valency of the counterion and whether the ion retains its hydration sheath when in the condensed state; one of the more recent attempts at the analysis is that of Satoh and Komiyama[51] – the intermediate model or IMM, a modified condensation theory. The model leads to interesting results when compared with experimental data for the titration of PAA, shown in figure 29; the smooth sigmoidal curves predicted by theory lead to an underestimation of pK_a at lower charge densities – considered by the authors to be due to limitations in the linear charge array model, adopted in condensation theories. The differences at higher charge densities is thought to be due to factors 'other than long-range' electrostatic interaction, since the model predicts saturation of the interaction due to counterion condensation.

Figure 30 illustrates the variation of activity coefficient of the counterion as a function of the charge density parameter ξ ($e/\epsilon kTb$, where e is the unit electronic charge, ϵ the dielectric constant of the bulk solvent, k the Boltzmann constant, T the Kelvin temperature and b is the axial charge spacing on the polyion). The empirical line describing the activity coefficient γ'_c is fitted to the equation

$$\gamma'_c = 0.96 - 0.42\xi^{\frac{1}{2}}$$

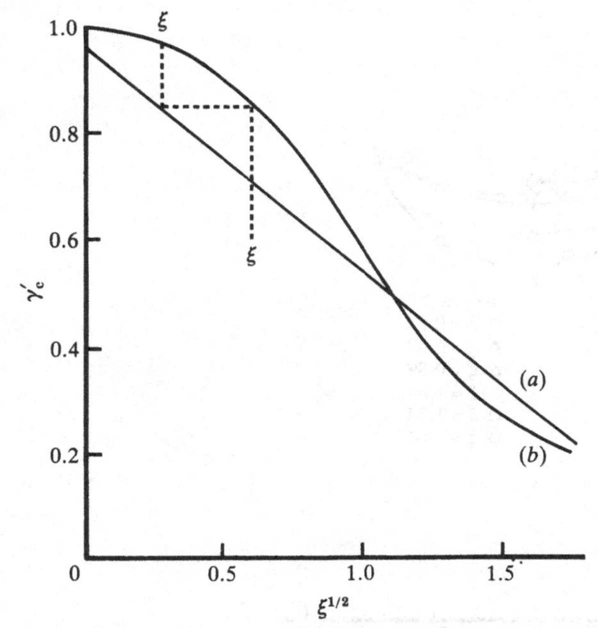

Figure 30. Dependence of experimental and theoretical counterion activity γ'_c upon the line charge density $\xi^{\frac{1}{2}}$: (*a*) empirical relationship of γ'_c and ξ for PAA/AAm copolymer; (*b*) IMM, $N = \infty$ $C_p = 0.0045$ eq dm^{-1} (data from ref. 51).

In the absence of added salt the theoretical values are consistently overestimated by the model at low charge densities, Satoh *et al.* suggesting that this is due to the polymer adopting a coiled conformation at these low charge densities; correction of γ'_c according to figure 30 results in the much improved fit of the dashed line in figure 29 at the lower charge densities.

Counterion–polyion interactions have also been studied by radio-tracer diffusion methods, most often involving the sodium ion[52], Figure 31 illustrates the results for aqueous salt-free solutions of NaCMC, over a polyion normality range of 10^{-4}–10^{-1} for various charge density values. The curves observed were similar to those reported for several different polyion sodium salts, heparin, alginic acid[53], *i*-carrigeenan[54], dextran sulphate, polyacrylate/acrylic acid[55], poly(styrene sulphonate) and poly(styrene carbonate)[56]. Fairly constant minimum values of $D_{Na^+}/D^{\circ}_{Na^+}$ (D_{Na^+}, the measured sodium ion self-diffusion coefficient, and $D^{\circ}_{Na^+}$, the diffusion coefficient in polyion-free salt solutions) were observed over the polymer concentration range N_p, where $10^{-3} < N_p < 10^{-2}$; at lower N_p values D_{Na^+} increased because the expanding ionic atmosphere gave increased screening of polyion charges and increased dissociation. D_{Na^+} increased at higher N values due to increased screening from overlapping ionic atmospheres between polyions. Figure 32 illustrates the dependence of $D_{Na^+}/D^{\circ}_{Na^+}$ upon

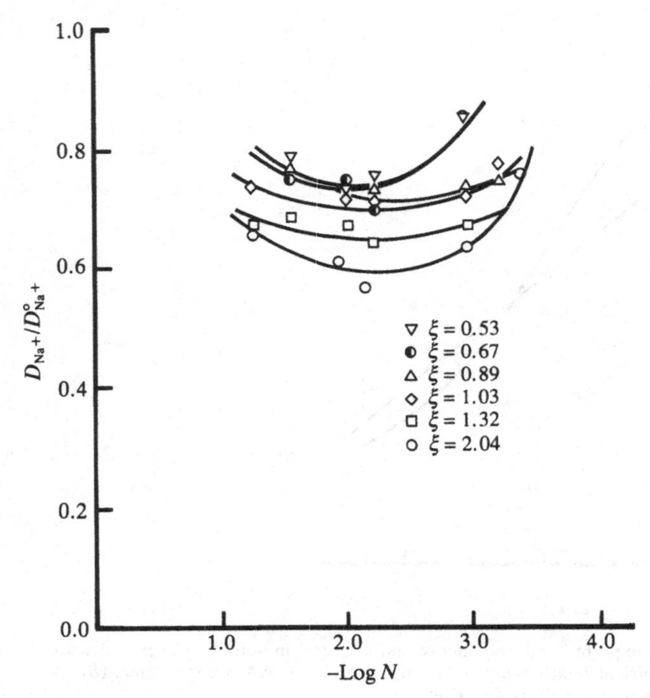

Figure 31. $D_{Na^+}/D^{\circ}_{Na^+}$ dependence upon the concentration of polyelectrolyte in salt-free solutions (data from ref. 52).

ξ^{-1} (predicted by both the Manning infinite line charge model and the cell model of Yoshida). When $\xi > \xi_c$ (ξ_c being the critical value of ξ below which, according to Manning, counterions are completely dissociated from the polyion and above which counterion condensation occurs until $\xi = \xi_c$) the linear dependence is observed for each polyelectrolyte studied. The average slope and intercept of all the vinylic polyelectrolytes (0.75 and 0.13) are nearer to the values expected from the Manning model, suggesting that the condensed sodium ions move with the polyion, which moves about two orders of magnitude slower than the counterions. In the case of NaCMC, however, the slope and intercept of NaCMC (0.32 and 0.41) are closer to the Yoshida predictions. This is possibly due to both the hydrophilic nature of the NaCMC and the rigid polymer backbone, compared to the hydrophobic surface and flexible backbone of vinylic polymers. In the region of $\xi > 1.32$, when the DS of the NaCMC is greater than unity, the value of ξ is probably large enough to make both kinds of polymer equally rigid. The hydrophilicity of NaCMC is of greater importance in this region, the attracted water molecules acting as a screen, weakening the long-range polyion-counterion interaction, resulting in larger diffusion coefficients. For $1.32 > \xi > 1.0$, the weaker electrostatic repulsion between the charged groups on the polymer

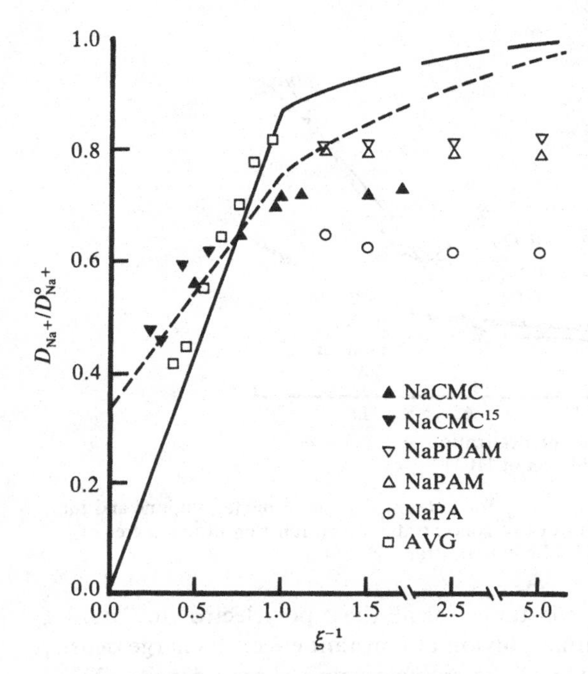

Figure 32. Minimum values of $D_{Na^+}/D^\circ_{Na^+}$ versus ξ^{-1} for NaCMC together with average values for NaPAM, NaPA and NaPDAM (data from ref. 52).

allows the chain to coil to a greater extent and the diffusion coefficient therefore rises. In consequence the sodium ions that interact with the more flexible vinylic polymers, have higher diffusion coefficients in this region; hence the smaller slope, when $\xi > 1$, for NaCMC polymers compared to vinylic polymers.

The rod-like models of both Manning and Yoshida predict that condensation will not occur when $\xi < 1$ for sodium polyelectrolytes and it is obvious from figure 32 that for NaCMC, when $\xi < 1$, a constant value of $D_{Na^+}/D^\circ_{Na^+}$ has been achieved. Extrapolation of the zero slope line to intersect with the experimental line when $\xi > 1$ yields an intersection close to $\xi = 1$, the ξ_c value – also observed with the other polyions reported. The data thus indicate that the critical charge parameter is correctly accounted for by theory.

Figure 32 also shows that when $\xi < \xi_c$ the values of $D_{Na^+}/D^\circ_{Na^+}$ are below those predicted by rod-like theories for all the polymers, suggesting the models are inappropriate within this range. Similar results have been obtained for sodium ion activity coefficients in sodium pectinate, NaCMC and NaPA/PAA. The zero slope line for NaCMC for $\xi < \xi_c$ is very similar to those for NaPA/PAM and NaPA/PDAM, also for $\xi < \xi_c$ the $D_{Na^+}/D^\circ_{Na^+}$

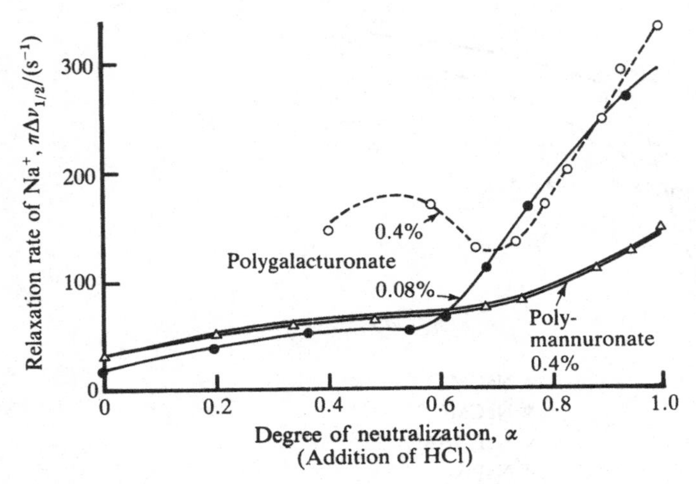

Figure 33. ^{23}Na nmr relaxation rate, $1/T_2$, for sodium polymannuronate solutions and for sodium polygalacturonate solutions at two concentrations as a function of the degree of neutralization, $T = 297$ K, freq. 26.4 MHz (data from ref. 58).

values for NaPA/PAA are constant. For all these polyelectrolytes below ξ_c the sodium ion interacts with a polyion of constant effective charge density, irrespective of the magnitude of the stoichiometric charge density. This is most likely to occur if the rod-like polyion folds when $\xi < \xi_c$ to a constant effective ξ value – possibly the position of minimum free energy of the solution. Klein and Ware[57] have found from electrophoretic mobility measurements of the 6–6 ionene bromide in solvents of different dielectric constants (and therefore different ξ values) that the mobility was constant for $0 < \xi < \xi_c$ but dropped abruptly at $\xi = \xi_c$; this behaviour suggests that the electrostatic field potential of the polyion remains constant below the value of ξ_c. It should be noted that this prediction of a coiled conformation below ξ_c is in agreement with the conclusions of Satoh *et al*, regarding the potentiometric titration of PAA.

Binding interactions by counterions have also been studied by nmr measurements using ^{23}Na. One of the more recent communications in this field is that of Grasdalen and Kvam[58], who reported on the binding of ^{23}Na by polyuronates and investigated the possibility of conformational changes; interpretation of the data on the basis of a single correlation time model for the sodium quadrupolar relaxation suggested a delocalized ion binding mechanism, as described by a simple Poisson–Boltzmann model, implying little specific ion binding in these natural polyelectrolytes. Of especial interest in this report are the data illustrated in figure 33, showing the effect of degree of neutralisation on the line width of ^{23}Na in solutions of sodium polymannuronate and sodium polygalacturonate, the latter at two different concentrations. For the most dilute polygalacturonate and polymannuronate solutions, even at low values of α, weak association of counterions is to be seen, but around $\alpha = 0.61$ and 0.72 for the respective polyions, where the line charge density exceeds Manning's critical value ξ_c, a rapid rise in counterion binding occurs. In the case of the more concentrated solution of polygalacturonate, however, a marked maximum is observed at $\alpha = 0.5$, indicating the onset of aggregation, the solution in fact becoming turbid.

The technique of potentiometric titration, previously mentioned in this review, is particularly valuable since the charge density of the polyion may be changed in a controlled manner, whilst the degree of 'condensation' may be independently varied by addition of electrolyte. When a polyacid bearing a large number of identical weak acidic groups is titrated with a base, the usual buffering capability of low molar mass analogues is missing, particularly at low ionic strengths. Since the weak polyacid becomes progressively charged during neutralisation, the Henderson–Hasselbach equation describing the dissociation of low molar mass acidic analogues must be modified to include an electrostatic potential energy term, accounting for the work which must be done in removing a proton from the increasingly electrostatically negative environment of the polyion, described by the equation[59]

$$pH = pK_a^\circ - \log[1-\alpha/\alpha] + 0.4343(dG_{el}/d\alpha)/RT$$

where pK_a° is the intrinsic dissociation constant of the ionised group in the polyion and $dG_{el}/d\alpha$, the electrostatic Gibbs energy change per unit degree of dissociation α.

The relationship of the apparent dissociation constant to the degree of dissociation α and the pH has also been described by the empirical relationship known as the Katchalsky–Spitnik equation

$$pH = pK_a - n \log[(1-\alpha)\alpha]$$

Unfortunately pK_a° and pK_a are not simply or directly related, Mathews[60] applied both these equations to the natural polyelectrolytes chondroitin-4-sulphate and dermatan sulphate, obtaining for chondroitin-4-sulphate the values $pK_a = 3.38$, $n = 1.32$, and $pK_a^\circ = 3.06$ and for dermatan sulphate $pK_a = 3.82$, $n = 1.25$ and $pK_a^\circ = 3.49$. The electrostatic Gibbs energy change per unit degree of dissociation, $dG_{el}/d\alpha$ is equivalent to the electrostatic energy term $\epsilon\psi_0/RT$; determination of ψ_0 values from electrophoretic mobility studies yields values twice as large as those calculated – these electrophoretic values would yield pK_a° values some 0.3 units lower than those calculated. The differences of 0.32 units between chondroitin and dermatan sulphates is accounted for by the different interactions between glucuronic carboxyl and sulphate groups and iduronic carboxyl and sulphate groups. Whilst no satisfactory correlation exists between pK_a° and pK_a, and in many publications this is ignored, it is reassuring to note that differences

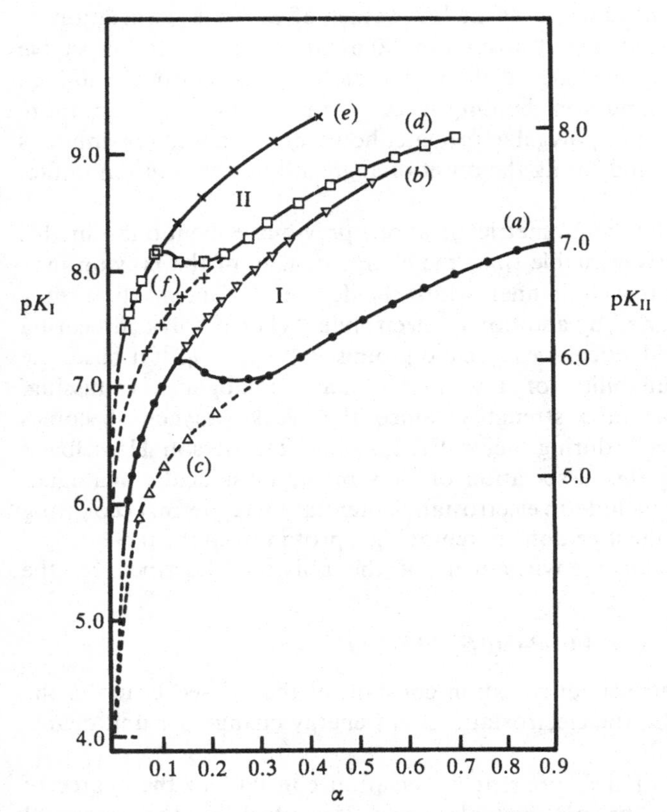

Figure 34. The apparent pK_a as a function of α for PMAA in aqueous solution. Curve I, 5 °C: (*a*), experimental data, (*b*) and (*c*) data from extrapolated Henderson–Hasselbach plot, above and below the conformational change. Curve II, 50 °C: (*d*), experimental data, (*e*) and (*f*) extrapolated as for 5 °C curve; (data from ref. 61).

Table 6. *Thermodynamic quantities for the conformational transition of PMAA derived from potentiometric titrations.*[a]

	278 K		323 K
ΔG	0.32		0.24
ΔH	0.54		0.46
ΔS		0.22	

[a] Units expressed per monomer mole kT^{-1}

between similar parameters for related macromolecules seems to remain the same.

Definitive early work by Mandel, Leyte and Stadhouder[61] produced interesting results on the conformational transition which occurs in PMAA as a function of its degree of dissociation, illustrated in figure 34; the associated thermodynamic parameters, shown in table 6, reveal that the enthalpies of conversion, together with the average value of the entropy of conversion, over the temperature range 278–323 K are all positive. The authors suggested that this did not provide evidence for hydrophobic bonding – this, however, was without consideration of the corresponding changes occurring in the water. Subsequent potentiometric titration data by Dubin and Strauss[62] on alternating copolymers of maleic acid and alkyl vinyl ethers, of increasing size of alkyl group, revealed that the methyl and ethyl copolymers behave as normal polyacids but the butyl and hexyl copolymers undergo a conformational change as α increases, from a hypercoiled state to a more extended conformation. The associated thermodynamic parameters show that ΔG_c passes through a maximum as a function of temperature, whose position is related to the size of the alkyl group; both ΔH_c and ΔS_c increase from negative to positive values with increasing temperature, strongly indicating the importance of hydrophobic effects in the conformational change. Direct calorimetric measurements of ΔH_c of PMAA by Crescenzi, Quadrifoglio and Delben,[63] produced a value of approximately 1 kJ mol^{-1}, compared with a value of 1.5 kJ mol^{-1} for ΔH_c obtained from interpolation of ΔG_c values at 25 °C and 45 °C – these data are in qualitative agreement with those of other workers, again showing the importance of hydrophobic interactions in the conformational change.

Further investigations by Strauss and coworkers involving calorimetry[64], solubilisation[65] and fluorescence[66] studies of maleic acid/alkyl vinyl ether copolymers have confirmed the importance of hydrophobic interactions in the conformational change. A recent reexamination by potentiometric titration of a series of maleic acid/alkyl vinyl ethers by Barbieri and Strauss[67] has led to the proposal that in the hypercoiled state (low α), each polymer molecule contains several small 'micelles', formed by the aggregation of adjacent hydrophobic units, rather like knots along a piece of string. More recently, Hsu and Strauss[68] reported fluorescence quenching measurements on the copolymer containing hexyl units as further support for this interpretation. Such a model certainly leads to interesting possibilities, but it is worrying that the recent work has been conducted in lithium chloride solutions, as distinct from the earlier work in sodium chloride; as the data in table 7 show, the nature of the counterion has an effect upon the acid dissociation constant of polyacids and the work of Nandi and Robinson[11] on the thermal denaturation of ribonuclease shows that differing counterions affect non-polar and polar groups to considerably different extents.

The microcalorimetric investigation by Daoust and Lajoie[69] of the heats

Table 7. *Apparent dissociation constants of acrylic and methacrylic acids as a function of cation type and concentration*

	PAA-pK_a		PMAA-pK_a	
Salt/Concn	1.0 M	0.1 M	1.0 M	0.1 M
Li^+	4.30	5.27	5.22	6.27
Na^+	4.65	5.30	5.35	6.37
K^+	4.83	5.37	5.60	6.44
Rb^+		5.40		6.50
Cs^+		5.52		6.58

of dilution of PMAA and its rubidium and sodium salts is particularly interesting for the information that it provides, both for the conformation of the polymer and the effects of differing ions; the data show that whilst the heat of dilution is exothermic over the concentration range investigated, the effect is much greater at the higher polymer concentrations – thought by the authors to be the consequence of hydrophobic hydration of the methyl groups previously buried within the interior of the macromolecular aggregate. Recent microcalorimetric measurements of the heat of dilution in the very water-rich region of PMAA[36] appear to confirm the very low values reported by the authors. The differing counterion effects upon the heat of dilution in the important higher concentration region are clearly seen, the dilution process becomes more endothermic with increasing concentration for the sodium salt, but passes through an exothermic minimum for the rubidium salt – evidence, according to the authors, that the sodium ion is much more strongly bound than the rubidium ion; evidence for the effect of the lithium ion would be especially interesting in this context.

Morcellet, Loucheux and Daoust[70] have also reported that further support for the role of hydrophobic effects in the conformational change of polyelectrolytes comes from their microcalorimetric investigation of the heats of dilution and heats of dissociation of the copolymer poly(N-methacryloyl-alanine-co-N-phenylmethacrylamide), P50, which also undergoes a conformational transition as α increases. Figure 35 illustrates the variation of the heat of dilution ΔH_d versus α for P50 and for the homopolymer poly(N-methacryloylalanine), PNMA and the model molecule N-isobutyrylalanine.

The sudden decrease in ΔH_d which commences at $\alpha = 0.30$ for PNMA coincides with the value of critical charge density for ion condensation, previously outlined; ΔH_d for P50, however, exhibits an endothermic peak over the same region, which appears to be related to the exposure of the hydrophobic groups of P50 to the water as a consequence of ionization.

The thermodynamic data for the transition of P50, obtained from microcalorimetric and potentiometric titration studies were; $\Delta H_d = 1.925$ kJ

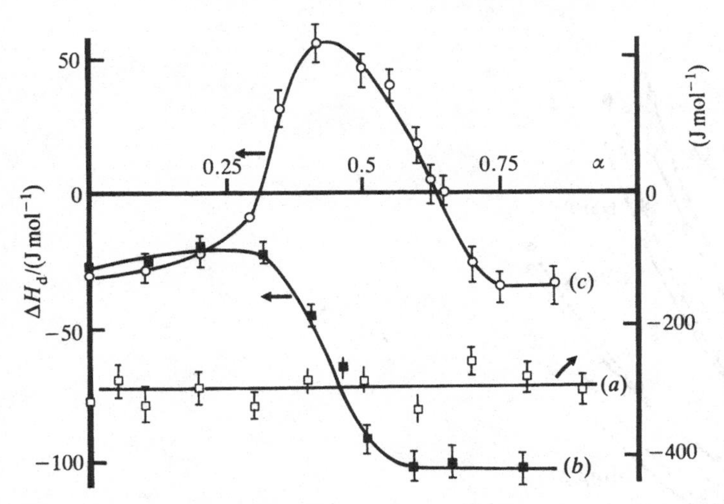

Figure 35. Variation of the heat of dilution ΔH_d vs the degree of dissociation for (*a*), NPAIB, (*b*) PNMA and (*c*), P-50 (data from ref. 70).

mol^{-1}, $\Delta G_d = 2.05$ kJ mol^{-1} and $\Delta S_d = -0.42$ J K^{-1} mol^{1-}; for the lower copolymers in the series (P38 and P24)[71], ΔS_d is approximately zero whilst for PMAA $\Delta S_d = 0.71$ J K^{-1} mol^{-1}, suggesting a decreased structuring of the final state. ΔH_d increases linearly with the amount of hydrophobic residues, again suggesting an increasing contribution from exposure of the nonpolar residues to the solvent.

A more extensive investigation of the effects of interaction of different ions with polyelectrolytes has been reported by van der Maarel, Lankhorst, de Bleijser and Leyte[72] who considered the dynamics of water molecules in solutions of PAA and PSSA and their alkali metal salts – ^{1}H and ^{17}O nmr data for longitudinal relaxation rates are shown in figure 36 for solutions of PAA and its salts: R_x° is the relaxation rate of the water molecule in the absence of polyelectrolyte and

$$R_x^\circ = fR_x^p + (1-f)R_x^u$$

where R_x^p and R_x^u are the relaxation rates of water perturbed by the polyelectrolyte and its condensed counterions, and of water associated with uncondensed counterions, f denotes the mol fraction of perturbed water, where

$$f = nc/55.5$$

c is the molality of the solution and n the hydration number. The data in figure 36 indicate that the water relaxation rates increase in the sequence

$$Li^+ > Na^+ > K^+ > Rb^+ = Cs^+$$

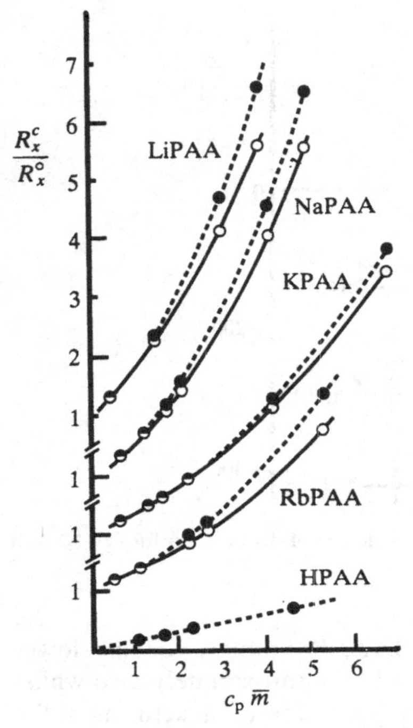

Figure 36. Isotope effect corrected longitudinal relaxation rates of fully neutralised solutions and PAA divided by the pure water values versus concentration: ^{17}O, filled symbols; ^{1}H, open symbols (data from ref. 72).

Recent neutron and X-ray scattering data on PMAA solutions[73] have shown the presence of a monomolecular hydration layer between the charged backbone and the condensed counterions, the density of the intermediate hydration water being approximately 8% greater than the value for bulk water. It appears that, due to the smaller cations having a stronger electrostrictive interaction with the polymer backbone, they accumulate closer to the backbone, inducing an increase in local water density in close proximity to the polyion – in consequence water reorientational mobility is reduced to a greater extent.

The difference in rates observed indicates that the overall effect upon the water molecules is due to both the polyion and the counterion, with the distance of closest approach being the important factor; this suggests that perturbation of water adjacent to the polyion surface extends for only one or two molecular layers.

Leyte and coworkers also investigated this possibility by the addition of excess of salts to the polyion solution; lithium chloride enhances relaxation as a linear function of lithium chloride concentration, whilst caesium chloride is virtually without effect. Qualitatively these results are similar to the influence of the salts upon water itself, without the presence of the polyelectrolyte. However, quantitatively, the rate of increase of relaxation due to lithium chloride is twice that of the caesium chloride solution, which confirms the enhanced effect of the condensed lithium ion counterion.

Nagata and Okamoto[74], however, have observed the reverse Hofmeister sequence; utilising rare earth metal fluorescence probe techniques, based upon the Tb^{3+} ion which shows fluorescence enhancement when bound to the polyion. NaPAA and NaPES excited considerable enhancement of the Tb^{3+} ion but NaPMAA and NaPSS did not. Titration of the Tb^{3+} ion solutions with Li^+, Na^+, K^+ and RbPAA solutions produced sharply increased fluorescence up to differing limiting values, from which binding constants for the various ions were derived; for the sequence Li^+, Na^+, K^+ and Rb^+ these were 300, 390, 500 and 600 respectively. Since the polyions of greater hydrophobic character showed no fluorescence enhancement, it appears that in this particular case the data reflect the relative difficulty experienced by the ions in displacing the Tb^{3+} ion from hydrophilic sites within the polyion, hence the reverse sequence.

The importance of hydrophobic interaction effects in the conformational change of PMAA and other polyacids is now clear, but other factors also have important contributions to the aqueous solution behaviour of polyelectrolytes, one of which, due to structural geometry is revealed by the investigations of Muroga and Nagasawa[75] of the potentiometric titration behaviour of PCA, the structural isomer of PMAA. Rotational motion around the main chain does not change the relative position of the geminal α-methyl group in PMAA, whereas the motion can change the relative positions of the vicinal β-methyl group in PCA. The β-methyl group can therefore intervene between adjacent carboxylate groups and affect the electrostatic interaction between them. Figure 37 shows the value of pK_a for isotactic and atactic PCA compared to atactic PMAA in low salt concentrations (0.01 M NaCl). There is no difference in the intrinsic acidity of the carboxylic acid groups in PCA and PMAA (pK_a, by extrapolation, being 4.7–4.8 for both acids) but the titration behaviour is distinctly different; no conformational change is seen in the case of PCA at low values of α, unlike PMAA, and in addition PCA is a much weaker acid at high values of α. It therefore appears that the conformation of PCA is a random coil, a conclusion supported by small angle X-ray scattering data[76], showing the persistence length of NaPCA (≈ 1.1 nm), to be comparable to those of randomly coiled vinyl polymers over a wide range of α. The insertion of the methyl group between the ionisable carboxylate groups in PCA seems therefore to have a two-fold effect, hydrophobic interaction at low values of α is prevented and the dissociation of the carboxylic acid groups at high values of α is suppressed.

The role of tacticity in the potentiometric titration behaviour of polyacids can be particularly informative; reference has been made to the two basic

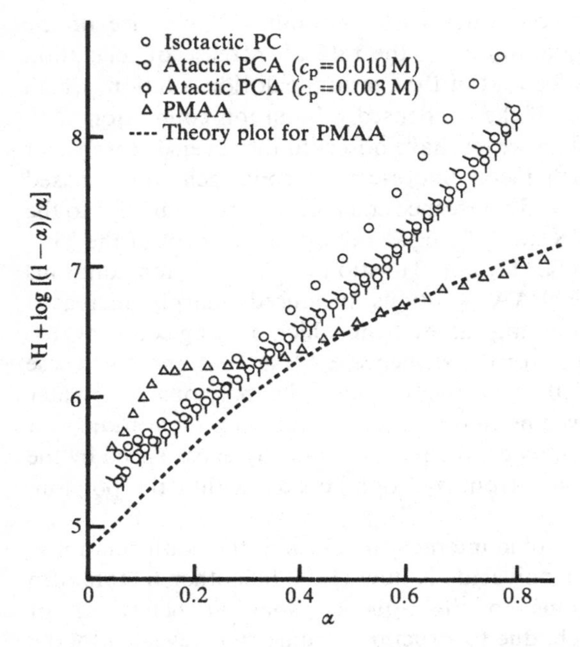

Figure 37. Potentiometric titration curves of PCA and PMAA at $C_s = 0.01$ M for PCA, isotactic PCA, atactic PCA, PMAA and the theory plot for PMAA (data from ref. 75).

models used in the analysis of the behaviour of polyions in aqueous solution and important differences are found between potentiometric titration curves calculated from the two different approaches. If the dissociation of the carboxylic acid group is determined by the average electrostatic potential within the polymer domain, several characteristics should be observed: (*a*) the data should depend on the molar mass of the polyion; (*b*) the potentiometric titration curves of dibasic acid copolymers, such as those of maleic acid, should not show the characteristic features of such acids; (*c*) the stereoregularity of the polyelectrolyte should have little effect on the potentiometric titration behaviour, since expansion of the polymer coil is little affected by stereoregularity; conversely, for the rod-like model, the opposite predictions arise[77].

Evidence has been published that the potentiometric titration curve of a polyelectrolyte is independent of the molecular mass[78] and the work of Strauss and others has shown that the titration curves of maleic acid copolymers show the features characteristic of dibasic acids – two items of experimental evidence in favour of the rod-like model; the data of Nagasawa, Murase and Kondo[79] on the titration of isotactic, atactic and syndiotactic PMAAs, shown in figure 38, indicate that the curve for isotactic PMAA always lies above that of the syndiotactic – the carboxylic acid groups in the

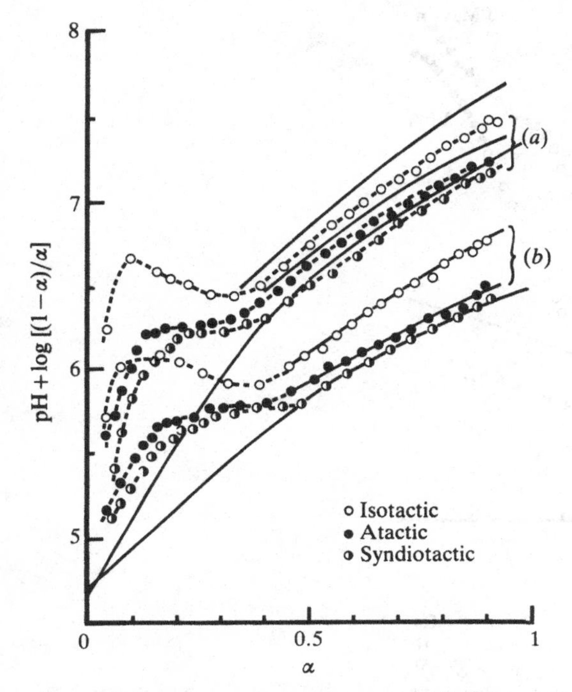

Figure 38. Potentiometric titration curves of PMAA at 23 °C, sodium chloride concentrations: (*a*) 0.010 M, (*b*) 0.10 M; for isotactic, atactic and syndiotactic forms of PMAA. The thin solid lines represent values extrapolated to infinite dilution (data from ref. 79).

isotactic form are closer together than in the syndiotactic, thus increased electrostatic work is required to remove the H^+ from the isotactic form. The curve for the atactic form always falls mid-way between the stereoregular forms of PMAA.

The good agreement of calculated curves of $0.4343/kT$ and the data for syndiotactic PMAA is reported by the authors as indicative of a planar zig-zag conformation for the syndiotactic form; the calculated curve of the isotactic form correlates with the experimental data when the charge density is determined on the basis of a helical coil of 3–4 monomers per turn.

More recently Chang, Muccio and St. Pierre[80] have determined the tacticity of PAA by ^{1}H and ^{13}C nmr to triad and partial pentad resolution, observing the change in the chemical shift of the methine resonance for each tacticity and of the carbonyl resonance with changing pH in a potentiometric titration. The effect of tacticity is small but the trend is clearly $pK_{mm} < pK_{mr} < pK_{rr}$ (m = meso and r = racemic); these data are, however, at variance with earlier results by Kawaguchi and Nagasawa[81], which suggested that the syndiotactic polymer, r > m, to be a slightly stronger acid than the isotactic form, m > r.

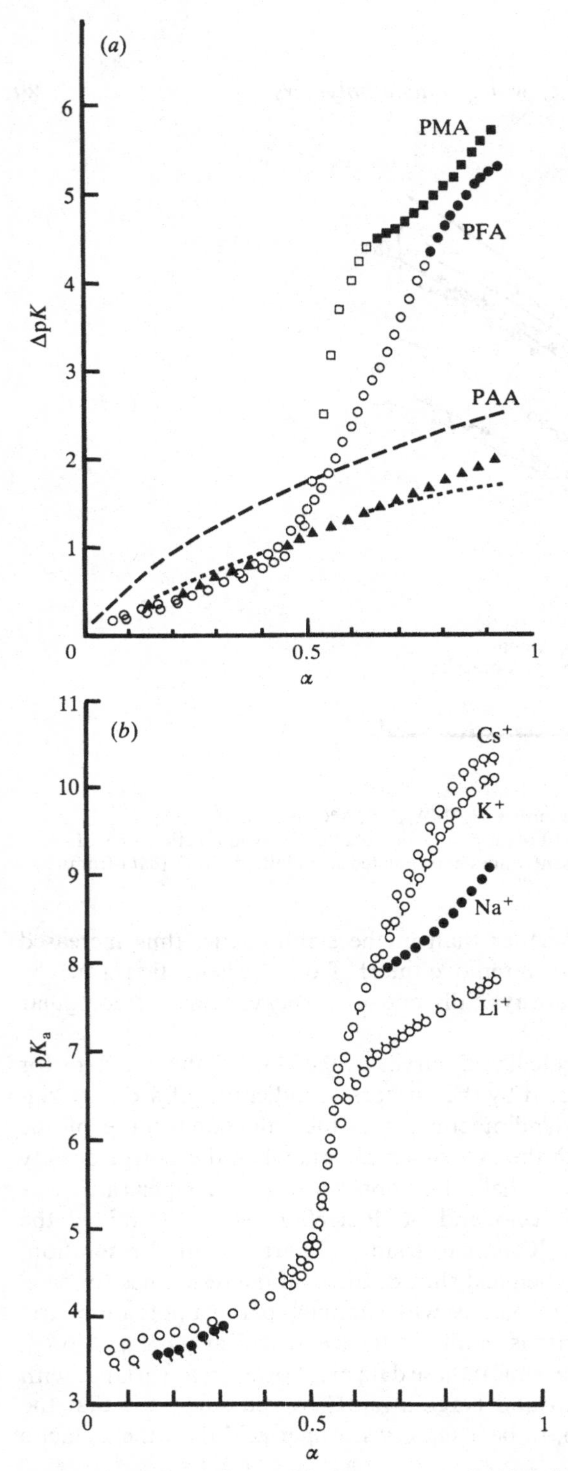

Figure 39. (*a*) Dependence of ΔpK on α at $c_s = 0.10$ M of PFA, PMA, and PAA. Filled symbols indicate turbid solutions. (*b*) Dependence of the apparent dissociation constant of PMA at $c_s = 0.10$ M upon the cation types: Li^+, Na^+, K^+, Cs^+. Filled circles indicate turbid solutions (data from ref. 82).

Table 8. *Intrinsic dissociation constants*

Acid	pK	pK
PMA	3.20	7.9
PFA	3.2	6.2
PEM	3.65	6.4

A more comprehensive insight into the detailed effects of molecular architecture of a polyelectrolyte upon its aqueous solution behaviour comes from the recent investigations of Kitano, Kawaguchi, Anazawa and Minakata[82] of the dissociation behaviour of the stereoisomeric polyacids, polyfumaric and polymaleic acids; figure 38(*a*) illustrates the behaviour of ΔpK, where

$$\Delta pK = pK_a - pK_0 = 0.4343(dG_{el}/d\alpha)/RT$$

at a constant salt concentration of 0.1 M for the two acids and PAA. At $\alpha < 0.5$ the 'first dissociation' of the polydibasic acids, assuming that the two carboxylic acid groups may be distinguished by their ease of dissociation, there appears to be little difference between all three polyacids in the electrostatic term, but the differences become very significant when $\alpha > 0.5$; ΔpK for PMA increases more steeply than that for PFA, that for PAA showing no change. In addition to these differences, at higher values of α both PMA and PFA solutions become turbid, PMA before PFA, indicated by the filled symbols in figure 39(*a*). The liability to turbidity is also very sensitive to the nature of the cation of the added salt, as also is the magnitude of ΔpK for $\alpha > 0.5$, there being a difference of three pH units in ΔpK between lithium ions, the lowest, and caesium ions; for $\alpha < 0.5$ only small differences in ΔpK exist but the ion dependency sequence appears to be reversed!

Since the large differences in ΔpK arise following the effective completion of the dissociation process for the 'stronger' of the two carboxylate groups, it would appear that these differences have their origin in the conformational aspects of the polyacids – PMA being more compact than PFA. Attempts to analyse the data in terms of a rod-like model were unsuccessful, even at low values of α and the data were analysed on the basis of two successive acid dissociations, as shown in table 8, indicating the second dissociation of PFA is considerably smaller than that of PMA, the electrostatic energy term is therefore greater in the case of PMA.

When added salts are present, the behaviour confirms the validity of separating the two dissociation processes – it should be noted that the data in figure 39(*b*) indicate that the turbidity occurs in the presence of caesium ion when $\alpha < 0.5$ but not when $\alpha > 0.5$, here turbidity is observed in the case of the sodium ions. Further detail is lacking in this intriguing situation but a localised salting-out appears to be occurring, temperature dependence data on this behaviour would be most instructive. Concomitant intrinsic viscosity data show that the second acid dissociation occurs with a decrease in the hydrodynamic volume of the polymer, thus chain expansion occurs during the dissociation of the first acid group but decreases on dissociation of the second. Kitano, Kawaguchi, Anazawo and Minakata[83] extended their investigations of the influence of molecular architecture upon the dissociation behaviour to include the alternating copolymer of isobutylene and maleic acid, hoping to clarify the relative efficiencies of the smeared and discrete charge models; PIMA is particularly useful in this respect since its chemical composition is identical to that of PMAA and PCA, although the two methyl and carboxyl groups in PIMA are situated closer to each other (the regularity of the polymer was extremely good, as checked by nmr). The dependence of ΔpK upon α showed that at low values ($\alpha < 0.5$) PIMA appears marginally weaker than PCA and PMAA, with no evidence of a conformational change such as is seen with PMAA; at higher values of α (> 0.5) a remarkable increase in ΔpK occurs, being highest of all for PIMA. The behaviour of ΔpK with respect to the nature of the added cation is similar to that of PMA and PFA, inasmuch that, below $\alpha = 0.5$ the sequence of values of ΔpK is $Li^+ < Na^+ < K^+$, whilst above $\alpha = 0.5$ the sequence is reversed; however, at high values of α the differences in respect of the ionic constituents are not as dramatic as in the case of PMA, nor do any regions of insolubility arise. It appears therefore that the dependence of ΔpK on the cation type increases as the overall charge density increases, but only above a lower limit of $\alpha = 0.5$; it has been suggested[84, 85] that this is due to the larger ions having less access to the polyion when the ionised groups are more densely arranged, and the greater state of hydration at these higher values of α. Such interpretations, however, do not seem to give sufficient weight to the influence of ionic hydration and changes that can occur during the 'condensation' process; this particular area of investigation would, it seems, benefit from detailed temperature dependent microcalorimetric studies such as those of Daoust and coworkers.

Analysis of the ΔpK data by the rod-like model, with smeared charge on the surface, satisfactory in the case of PAA and PMAA at high values of α, is unsatisfactory in the case of PIMA. However, analysis of the data to include explicit short-range interaction in the model, resulting from discrete charge centres, rather than smeared charge, via the Ising model[86], produced essentially two contributions to ΔpK, as follows

$$\Delta pK = \Delta pK_{\text{short}} + \Delta pK_{\text{long}}$$

where ΔpK_{short} is due to short-range interaction, up to the second nearest neighbour and ΔpK_{long} is due to that beyond. The resulting ΔpK values, together with those of ΔpK_{short} and ΔpK_{long}, are shown in figure 40, assuming a dielectric constant for solvent water of 78; assumption of a lower dielectric constant (≈ 15) only in the vicinity of the carboxyl group resulted in a

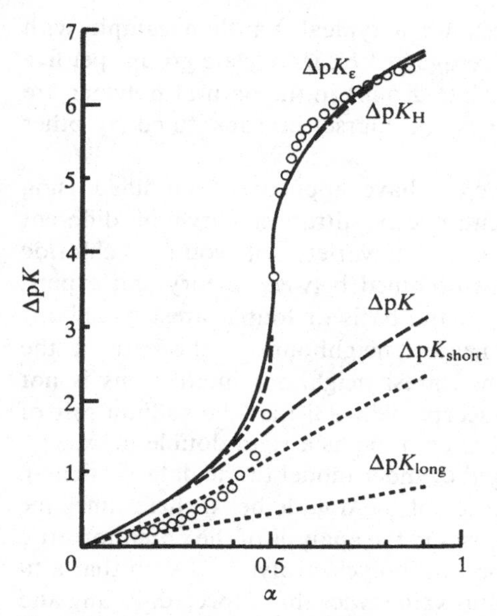

Figure 40. Comparison of calculation based upon Ising model with experimental ΔpK_0 of PIM. Calculated lines for: ΔpK without hydrogen bond at $\epsilon = 78$ (broken line); $\Delta pK(\epsilon)$ without hydrogen bond at $\epsilon = 15$ in nearest neighbours (solid line), ΔpK_H with hydrogen bond between nearest neighbours at $\epsilon = 78$ (chain line). Dotted lines are ΔpK_{short} and ΔpK_{long} for model calculation without a hydrogen bond at $\epsilon = 78$, with ΔpK represented by the broken line (data from ref. 86).

considerably improved fit to the data (ΔpK_ϵ). Ultraviolet and infrared spectra suggest that hydrogen bonding between carboxyl groups can also occur. Allowance for this, at an unchanged value for the dielectric constant of 78, results in the line ΔpK_H; it would seem that, since both effects are likely to be present, the fit of the model is a reasonable explanation of ΔpK behaviour at high α values, but clearly not when α is low, where the effects are largely long range in nature.

Interest in the area of natural and semi-synthetic polymer electrolytes has centred recently on the commercially important polysaccharide Xanthan gum – this polymer has a linear cellulosic backbone consisting of 1–4 *β-D*-glucose residues, with a three sugar substituent group attached at C-3 of every second backbone glucose unit, producing a regular repeating copolymer with comblike branching. The proximal *α-D*-mannose residue is normally acetylated at C-6 and the distal *β-D*-mannose may carry a pyruvic acid residue in a ketal linkage at C-4 and C-6; the degrees of acetyl and pyruvyl substitution DS_{ac} and DS_{pyr} are defined as the fraction of the respective sites substituted, the degree of pyruvation typically varying between 0.31 and 0.60. In addition to the carboxyl group present on each pyruvyl-substituted side chain, every side chain carries a carboxyl group on the central *β-D*-glucuronic acid residue; a typical Xanthan sample with $DS_{pyr} = 0.6$ will therefore carry an average of 1.6 carboxylate groups per five repeating sugar units. These carboxylate groups in the natural polymer are associated with sodium ions, but may, of course, be substituted by other singly charged counterions.

Zhang, Takernatsu and Norisuye[87] have applied the modified Ising model of Matsumura to the potentiometric titration curve of different molecular mass Xanthan samples at a variety of sodium chloride concentrations; good agreement was obtained between theory and experiment at higher salt concentrations on the basis of four nearest neighbour interactions, rather than the two nearest neighbours in the case of the copolymer PIMA; the difference in nearest neighbour interactions is not unexpected, since it is now widely accepted[88–93] that the sodium salt of Xanthan dissolves in aqueous sodium chloride as a rigid double helix.

Attempts to fit a uniformly charged cylinder model to the data yielded an unreasonably large value for the radius of the double helix, three times the expected value. A similar situation arises in the analysis of the potentiometric data of the single strand α-helical natural polyelectrolyte PGA, in that a fit of the data is only achieved for a radius value twice that expected. Zhang and coworkers have shown that the modified Ising model, with seven nearest neighbour interactions, is a reasonable fit over the range of α where the helical conformation is known to exist.

It has been shown by a variety of physicochemical techniques that Xanthan undergoes a rather diffuse conformational change with increasing temperature[94–9] and the recent light-scattering investigation by Hacche, Washington and Brant[100] of the change suggests that, up to a temperature of approximately 70 °C, the equilibrium system contains partially dissociated double strand dimers, in addition to aggregates comprising more than two chains. The data additionally suggest that, in the case of low ionic strength, where the Donnan contribution to the second virial coefficient is dominant, partial dissociation of the double stranded component leads to a large increase in the magnitude of the second virial coefficient with increased temperature.

5. Interpolymer complex formation

Complex formation between water-soluble polymers involves either charge interaction between oppositely charged polyions, or hydrogen bonding, involving the undissociated or partially dissociated form of the polyacid or base and a nonionic polymer; in addition to charge interactions and hydrogen bonding other factors, such as hydrophobic interaction will also have a role to play in complex formation.

Interpolymer complex formation may occur by one of two major routes:

(1) From pre-existing macromolecules, with some structural and chemical group similarities, by a straightforward mixing process.
(2) Polymerisation of a monomer in the presence of a second macromolecular component, known as matrix polymerisation.

In the first case complex formation ensues from random contacts, whereas in the second case, a 'zip-up' type of mechanism leads to the double stranded complex.

The different mechanisms lead to different structures and properties for the complex, when formed by the same two polymers, by the different methods; for example, complexes of PMAA and PDMAM obtained by mixing and matrix polymerisation of PDMAM in the presence of PMAA show different compositions – in the first case the base molar ratio (the usual units in which complex composition is expressed) is 3:2, whereas in the second case the ratio is equimolar.

5.1. *Polyelectrolyte complexes*

The formation of a polyelectrolyte complex is controlled by the physicochemical characteristics of the individual components of the complex – the strength of the dissociating acidic or basic groups, their position in the macromolecular backbone and the extent of their dissociation; in addition the physicochemical state of the aqueous solvent will be of importance, such factors as temperature, pH, ionic strength and the nature of the ions present in the solution (a quality-of-solvent effect). The complex, after formation, may remain in solution or separate as a gel or a solid precipitate, depending on variations in the factors discussed.

A good illustration of the importance of the architecture of the macromolecules in the formation of a polyelectrolyte complex is provided by the investigations of Tsuchida, Osada and Sanada[101] of the interaction of NaPSS with QPEVP and the ionene 3X; from the structures shown in diagram 3 it is clear that in the case of QPEVP the cationic group is pendant to the macromolecular backbone, whilst in 3X the cationic centre is integral within the chain. The structure of the insoluble complex of QPEVP and NaPSS is 1:1 and unaffected whether QPEVP is added to NaPSS or vice

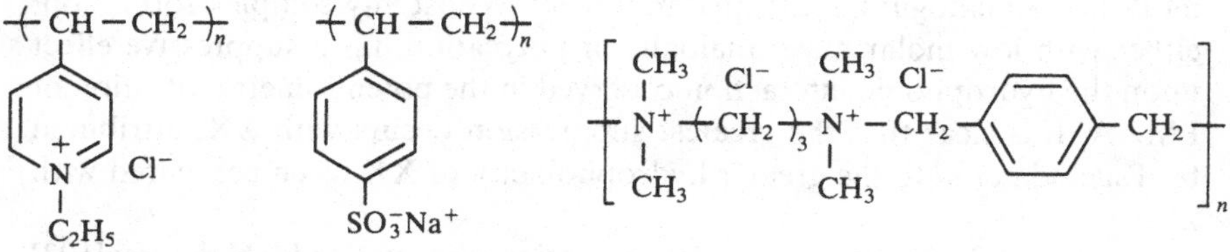

QPEVP NaPSS 3X

Diagram 3

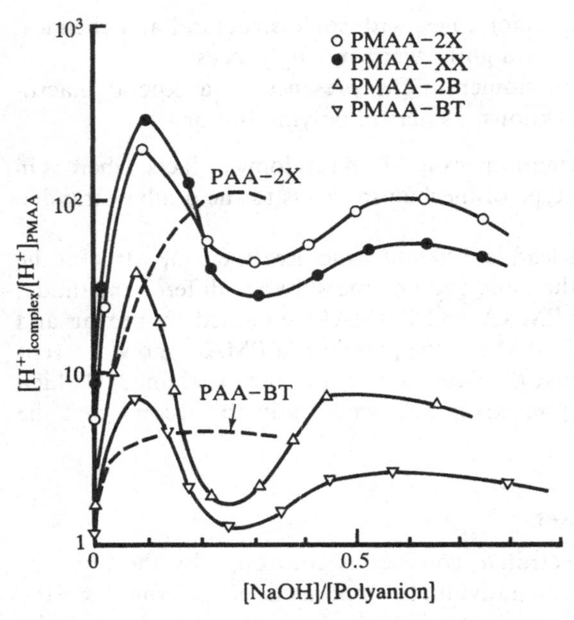

Figure 41. Amount of protons released from PMAA during formation of complexes, as a function of degree of dissociation (data from ref. 102).

versa; in the case of the complex of NaPSS and 3X, however, this is not so, addition of 3X to NaPSS results in an insoluble 1:1 complex, but addition of NaPSS to 3X first of all results in the insoluble 1:1 complex, but continued addition of NaPSS results in the formation of a soluble 3:1 complex, the ionene cationic site being open to further attack by excess NaPSS.

Figure 41, from the data of Tsuchida and Abe[102], illustrates the increased extent of dissociation of PMAA in the presence of the polycations 2X and XX, together with their low molar mass analogues BT and 2B (diagram 4) when compared with PMAA alone, during the course of potentiometric titration; also indicated is the increase observed for PAA—2X and PAA—BT. The presence of the polycations greatly increases the extent of dissociation of both PMAA and PAA as compared to the low molar mass analogues, illustrated in table 9. Whilst any complex formation, either with low molar mass analogue or polycation has a suppressive effect upon the hydrophobic interaction observed in the potentiometric titration of PMAA, it is clear that the greatest suppression occurs with XX, attributed by Tsuchida *et al* to the greater hydrophobicity of XX when compared with 2X.

The effect of tacticity on complex formation was studied by Nakajima[103] in PLL complexes with isotactic, syndiotactic, conventional and atactic PMAA, it was found that the helix content of PLL in the complex varies with

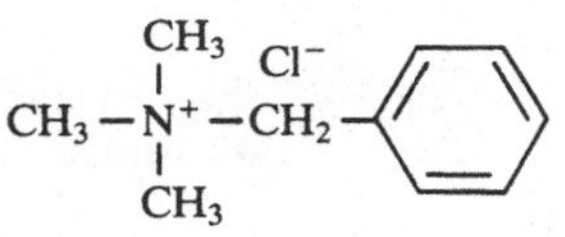

BT

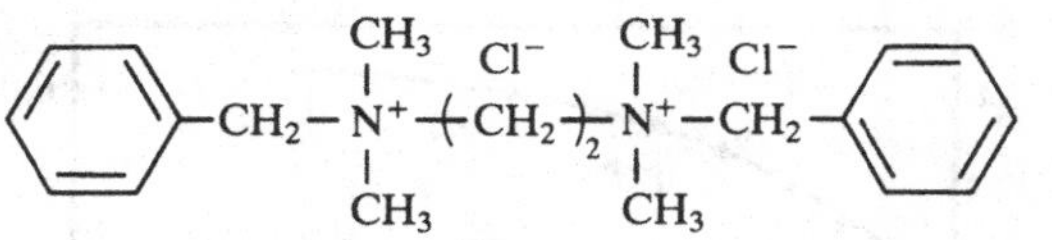

2B

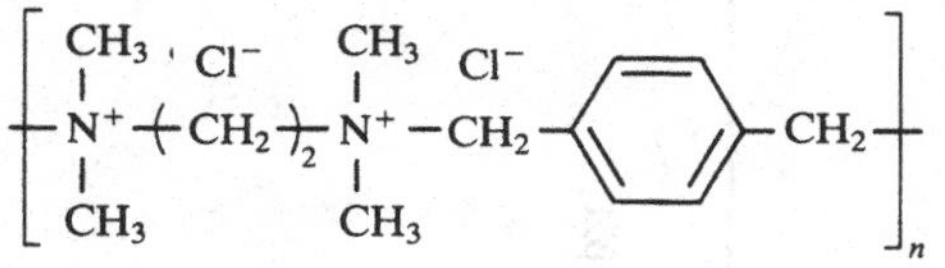

2X

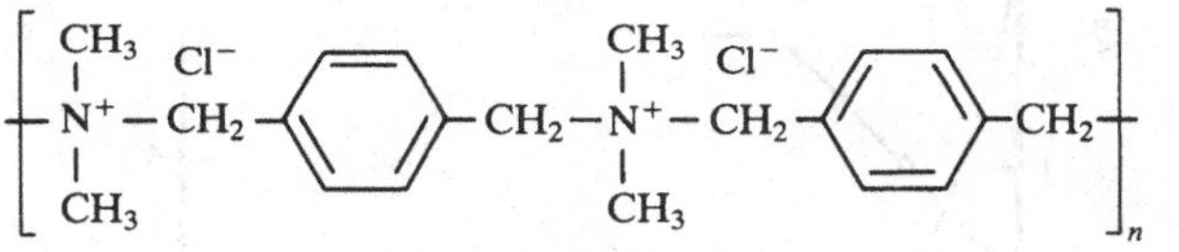

XX

Diagram 4

Table 9. *Intrinsic dissociation constants of PMAA and its complexes*

Sample	pK	*n*
PMAA	7.3	2.3
PMAA—BT	6.3	2.2
PMAA—2B	6.4	2.2
PMAA—2X	4.3	1.4
PMAA—XX	4.8	

the PMAA configuration – isotactic PMAA induces an α-helical structure, whereas conventional PMAA with a lower regularity of structure shows successively an enhancing and disruptive effect with increasing PMAA content. Nakajima reported that the helix disruptive effect is proportional to the atactic content of PMAA in the sequence:

iso $\ll$ syn $\ll$ conventional $<$ atactic

The influence of solvent characteristics upon polyelectrolyte complex formation has also been reported by Tsuchida and Abe[102] from an examination of the effect of ionic strength upon the viscosity and transmittance of the aqueous solution of the PMAA—2X complex. The data showed that, in contrast to the NaPSS—2X complex, for α values greater than 0.5, dissociation into the component polyelectrolytes occurs at sodium chloride concentration of 0.7 M, independent of complex composition – due

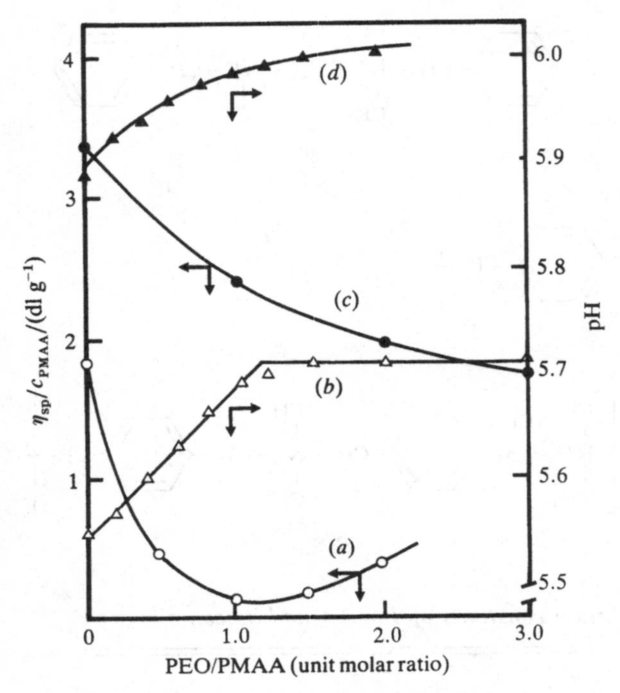

Figure 42. pH and viscosity changes for a base mole unit ratio of PMAA/PEO: (*a*), (*b*) initial pH 5.5; (*c*), (*d*) initial pH 5.9; $T = 25$ °C (data from ref. 104).

to the screening effect of the low molar mass ions. At $\alpha < 0.5$, however, the complex precipitates intact with increasing ionic strength, due presumably to the increased extent of hydrogen bonding at these lower levels of dissociation; precipitation therefore appears to be a quality-of-solvent effect, perhaps, in the light of evidence reviewed, accompanied by a conformational change.

5.2. *Complex formation involving hydrogen bond formation*

Complexes of this kind are formed between proton accepting and proton donating polymers, the most common being between synthetic polymers, involving weak polycarboxylic acids, such as PMAA and PAA, with nonionic polymers such as PVOH, PEO and PVP. Figure 42, from the data of Ikawa, Abe, Honda and Tsuchida[104], illustrates the importance of pH in this kind of complex formation; at an initial mixing pH of 5.9 no complex appears to be formed, whereas at an initial mixing pH of 5.5 complex formation apparently occurs at a PEO:PMAA ratio of 1:1.

The authors suggest that with too high a degree of dissociation of the carboxylic acid groups (above a 'critical pH') the number of active sites for hydrogen bond formation is insufficient to stabilise formation of the

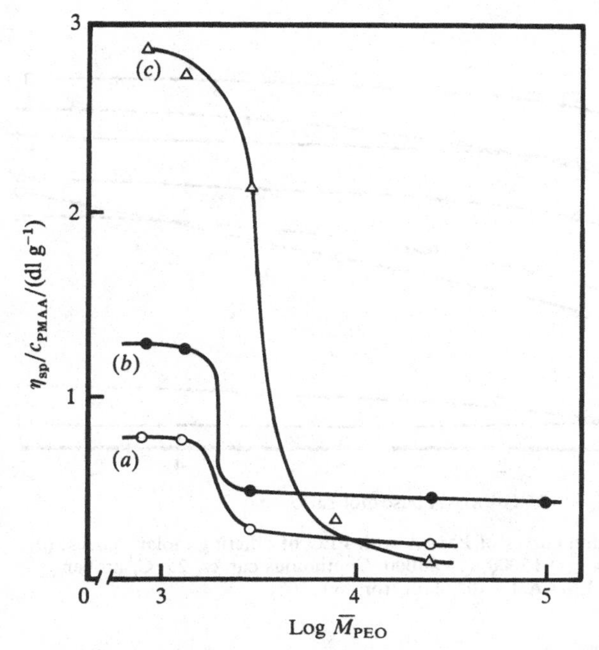

Figure 43. Dependence of the reduced viscosity of the PMAA/PEO system upon the PEO concentration M_{PEO}: (*a*) water; (*b*) 25% w/v methanol; (*c*), 50% w/v methanol. $\bar{M}_n$ (PMAA) = 147000, $\bar{M}_w$(PEO) = 1.4×10^5 (data from ref. 104).

complex; the critical pH value of PMAA is about 5.7, approximately 5.2 for the alternating copolymer PSMA and about 4.8 for PAA. The 'critical values' of pH thus follow the trend of the corresponding pK_a values, 7.3, 6.5 and 5.6 respectively; from the date of Tsuchida *et al* it appears that complex formation involving maleic acid copolymers may involve only the stronger of the two carboxylic acid groups; however, this remains open to question at the present time since little investigation of complexation involving dibasic polyacids seems to have been reported.

The effect of hydrophobic interactions upon complex formation was also considered by these authors, together with the effect of critical chain length of the PEO; figure 43 illustrates the effect of methanol concentration upon the reduced viscosity of PMAA/PEO in methanol/water mixtures (for a 1:1 base mol ratio in the complex). The sharp reduction in η_{sp}/c observed over a narrow range of molar mass of PEO is due, according to the authors, to a cooperative effect, smaller molar mass of PEO (below about 2000) precluding formation of the complex. The largest reduction in η_{sp}/c, observed at 50 wt% methanol concentration, coincides with the maximal effect of this methanol concentration upon the structure of solvent water, leading to a reinforcement of the hydrophobic interaction effect.

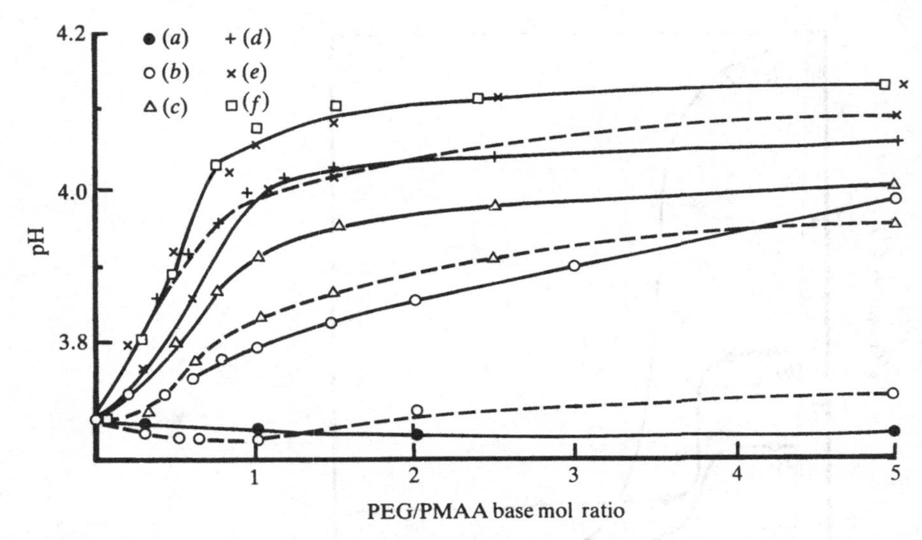

Figure 44. Potentiometric titration curves of PMAA with PEO of differing molar masses; (*a*) 1000, (*b*) 2000, (*c*) 3000 (*d*) 6000, (*e*) 15000, (*f*) 40000. Continuous curves, 25 °C, broken curves, 16.5 °C. Molar mass of PMAA 1×10^5 (data from ref. 108).

A similar maximal solvent effect is found in the studies by Bimendina, Saltybaeva and Bekturov[105] of the influence of ethanol concentration in ethanol/water mixtures upon the values of η_{sp}/c as a function of the degree of ionisation of PVP—MAA/MMA complexes; the maximum effect observed at an ethanol concentration of 30 wt% corresponds closely with the effect of this ethanol concentration upon solvent water structure[106].

Further evidence for the subtle nature of the interaction in complex formation is obtained from the potentiometric and conductometric investigations of Bekturov, Bimendina and Ileubaeva[107] of the interaction between a copolymer of VP/AA(70/30) and PAA or PMAA; no complex formation occurs in the case of PAA but clear evidence of complex formation occurs with PMAA, indications that interaction can be highly selective in character.

Complex formation in dilute solution is generally considered to be stoichiometric in the ratio of 1:1 per base mol of proton donor to proton acceptor and much of the literature describing effects such as hydrophobic interaction, ionic strength etc. with regard to complex formation, refers to a 1:1 base mol ratio in the complex. It should, however, be pointed out that the great majority of data is obtained from potentiometric, conductometric or viscosity studies, those of Antipina, Baranorskii, Papisov and Kabanov[108] being a good illustrative example. Figure 44 taken from this work, illustrates the titration curves of PMAA with PEO of differing molar masses (at 25 °C and 16.5 °C). The rise in pH as the polyacid is titrated with PEO

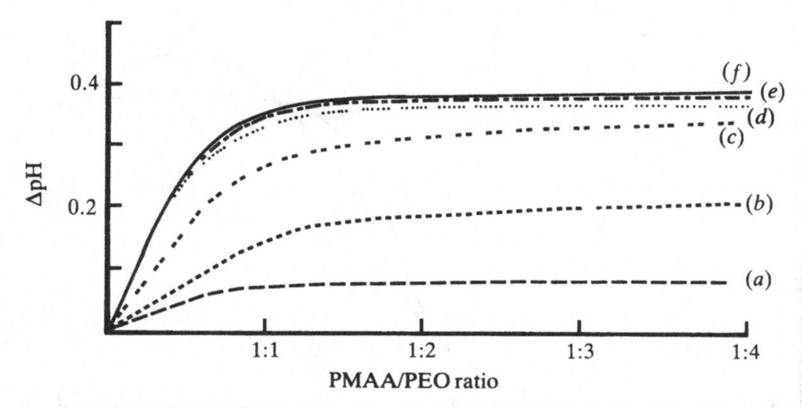

Figure 45. ΔpH for difference potentiometric titration curves of PMAA (14000), with PEO of varying molar mass against water: (*a*) 1000; (*b*) 1500; (*c*) 2000; (*d*) 6000; (*e*) 20000; (*f*) 35000 (data from ref. 36).

is taken as evidence for complex formation, with no complex being formed in the case of the lowest molar mass of 1000; the maximum value of pH achieved at a PEO/polyacid ratio of 1:1 is taken as evidence of complex formation. Corresponding reduced viscosity data indicate a minimum value around the same base mol ratio, also taken as evidence for the 1:1 complex – note, however, that the results at 16.5 °C are much less convincing in this respect!

More recently similar titration procedures, involving a greater spread of molar mass of PEO, provided the results shown in figure 45. The same general pattern is observed, but evidence for some complex formation is discernible with the PEO of molar mass 1000; the values of ΔpH clearly tend to a plateau value for base mol ratios greater than 1:1 – when the molar mass of the PEO is greater than about 10000; this is also the case with the data of Antipina and others.

Direct flow microcalorimetric measurements of the heat of complex formation indicate that no maximum value for the heat of complex formation is seen when the mol ratio is 1:1; the heat continues to rise with increasing PEO concentration; a maximum value of ΔH, is, however, observed as a function of molar mass, at approximately 10000, shown in figure 46 – the same value of molar mass at which ΔpH achieves a maximum value. Corresponding apparent partial molal volumes of PMAA in the complex also show no maximum as a function of the base mol ratio, but do show a maximum value as a function of molar mass, at a value of 10000. An interesting correlation exists between these data and the heat of dilution of PEO data as a function of molar mass, reviewed earlier: the heat of dilution of PEO decreases with increasing molar mass and achieves a limiting value at the same value of molar mass at which the heat of complexation passes

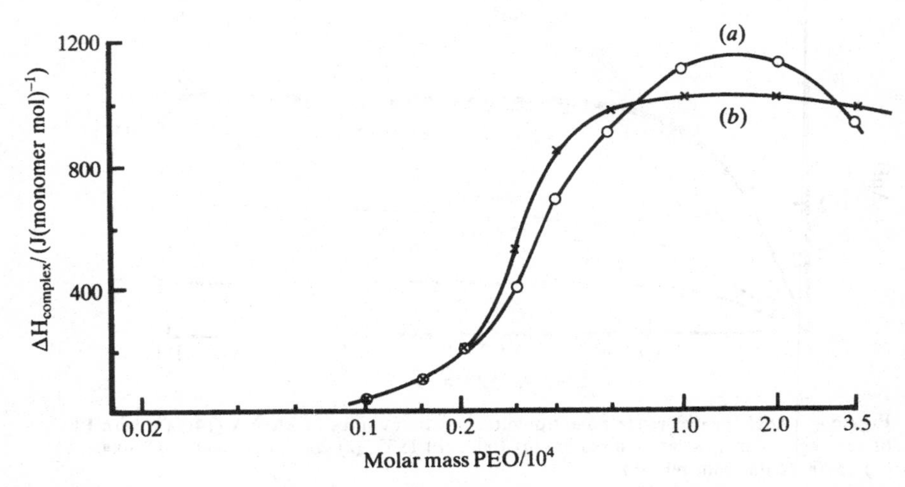

Figure 46. Heat of complexation data for PMAA/PEO, at base mol ratio of unity: (*a*) PMAA $\bar{M}_w = 44000$; (*b*) PMAA $\bar{M}_w = 14000$, $T = 25$ °C (data from ref. 36).

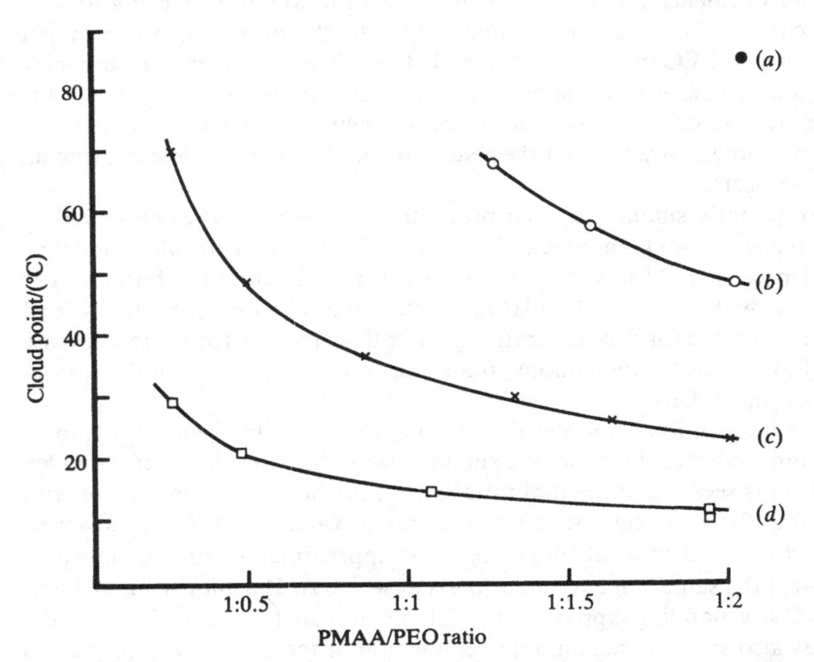

Figure 47. Cloud point data for PMAA/PEO complex formation as a function of PMAA/PEO ratio and PEO molar mass: (*a*) 400; (*b*) 600; (*c*) 1000; (*d*) 1500. PMAA concentration 0.8% w/v (data from ref. 36).

through a maximum. Figure 46 illustrates data obtained with PMAA of 14000 and 40000; it is clear from these results that the effect is largely independent of the molar mass of the PMAA and is a function of the molar mass of the PEO.

Measurements of the cloud points of PMAA/PEO complexes, shown in figure 47, for a range of PMAA/PEO ratios and a range of PEO molar masses, indicate that when the molar mass of PEO exceeds 1500, the insoluble complex is observed at all temperatures between 0 °C an 100 °C, but for lower PEO molar masses the cloud point becomes progressively higher. Several points of interest arise: firstly, there is no obvious discontinuity at a molar ratio of 1:1; secondly, it is clear from these data, together with those obtained from microcalorimetric studies, that soluble complexes are formed at lower temperatures and PEO molar masses; and thirdly, the LCST behaviour of the complexes is strongly indicative of a hydrophobic contribution to the solution behaviour of the complex.

Osada and Sato[109] earlier reported the effects of temperature on the potentiometric titration of PMAA with PEO of molar masses ranging from 200–20000. They measured the fraction of carboxyl groups associated with ether oxygens, y; the points at which these workers detected measurable values of y seem to correspond with the temperatures at which cloud points are observed. Calorimetric data, however, clearly show the existence of soluble complexes of lower PEO molar masses, indicating a certain lack of sensitivity in the potentiometric technique; the data of Osada and Sato do, however, reveal the existence of a maximum value of y somewhere between PEO molar masses of 7500 and 20000, which is in broad agreement with the evidence of a maximum effect associated with a PEO molar mass of 10000, as derived from other data.

It would seem, from the information considered, that techniques based upon monitoring the existence of hydrogen ions suggest that PMAA or PAA form complexes with PEO in a 1:1 ratio, whilst techniques such as calorimetry or density measurements give no such indication. The data of Bednar, Huang, Chang and Morawetz[110], obtained from fluorescence studies of complex formation between dansyl-labelled PAA and PEO, are particularly interesting in this respect, since fluorescence enhancement showed no discontinuity at a PEO/PAA ratio of 1:1. In fact, a continuous change was observed up to a ratio of 32:1. The data also suggest that for lower PEO molar masses the emission maximum undergoes a blue shift, whilst for higher molar masses the emission maximum undergoes a red shift. The authors suggest that, for lower PEO molar masses the dansyl label is effectively removed from contact with water if excess PEO is present. For higher molar mass PEO the Dan—PAA chain is in contact with the PEO only at widely separated regions along the contour of the chain, which is stretched out between these contact points. This interpretation is very much at variance with that previously reported for the 1:1 complex[111], that long uninterrupted sequences of bonds exist in polycomplexes, via such structures

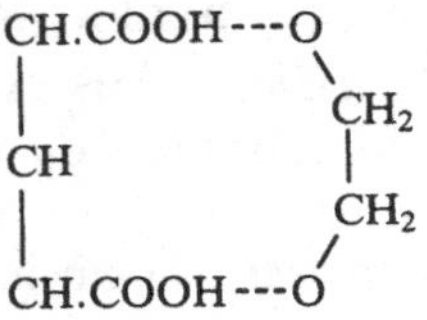

```
CH.COOH---O
|          \
|           CH2
CH          |
|           CH2
|          /
CH.COOH---O
```

Diagram 5

as that shown in diagram 5 which do in fact require the existence of 11 membered rings! Confirmation of the lack of such structures comes from the introduction of 9 mol% acrylamide into PAA – complexation is unaffected although PAAm is known not to complex with PEO.

Oyama, Tang and Frank[112, 113] have studied the pyrene excimer fluorescence of pyrene end-labelled PEO in complexation with PAA and PMAA; these workers again observed continuous change in excimer fluorescence as a function of PEO/polyacid ratio; no evidence for a 1:1 complex was established – the investigators did, however, establish that in the case of PMAA the mobility of the PEO chains is much more restricted than is the case with PAA, a reflection of the hydrophobic interaction effect of the α-methyl group.

The indications of aggregate formation by PEO of molar mass in excess of $(1–2) \times 10^4$, discussed previously, leads to the possibility that the divergencies reported here are a reflection of the change in PEO structure in the formation of the complex. It should be noted that, in the case of complexation of Dan—PAA with PVP however, the complex is consistent with the assumption that it contains equivalent concentrations of hydrogen bond donor and acceptor groups[14].

Further to the earlier comments upon evidence for 1:1 stoichiometry, derived from potentiometric titration data, the recent paper by Iliopoulos and Audebert[115] is of particular relevance. These authors report potentiometric measurements of complex formation between PEO, PVME and PVP with partially neutralised PAA; potentiometric measurements, however, were not undertaken during a titration. Instead, the components of the complex were mixed and allowed to stand for one hour prior to measurement of the pH – it was subsequently found that complexation increased continuously with the concentration ratio polymer/PAA, with no evidence of 1:1 stoichiometry. Increasing the degree of dissociation α resulted in a decreased extent of complex formation, attributed by the authors to a reduction in the length of uninterrupted hydrogen bond sequences. This is, of course, in contradiction to the fluorescence evidence of Morawetz and coworkers, but in this case the investigators were effectively introducing charged groups into the macromolecule which must be reflected in the conformational flexibility of the polymer backbone.

Thermodynamic data are often most helpful in determining the role of hydrophobic interaction in an equilibrium process and should therefore be of value in examination of the complexation process. A direct calorimetric measurement of the heat of complex formation by Papisov *et al*[116], yielded a value independent of molar mass, using 15000 and 40000 PEO, of approximately 1300 J mol^{-1} for a complex with 1:1 ratio of PMAA/PEO, but decreasing to 1100 J mol^{-1} for 6000 PEO – a trend similar to that shown in figure 46. The values for the heat of complexation using a 20000 PEO in 30% methanol/water solvent was −700 J mol^{-1}, thus the process is endothermic in water but exothermic in methanol solution; corresponding values for PAA/40000 PEO were 550 J mol^{-1} and −750 J mol^{-1} respectively. The authors claim the value for PMAA/PEO is in satisfactory agreement with a value of 1400 J mol^{-1} for ΔH obtained from potentiometric titration of the complex; it should be noted, however, that the temperature dependence of pK versus α is not a smooth progression as a function of temperature; examination of the data shows that the values for 10 °C lie between those for 25 °C and 40 °C – the derived values of ΔG and therefore ΔH should, in consequence be treated with some caution.

Calorimetric investigations by Tsuchida and Abe[117] of the formation of the PMAA/PEO complex, involving high molar mass polymers in a 1:1 ratio produced a value of 1300 J mol^{-1} in water at 25 °C, in good agreement with the calorimetric value obtained by Papisov *et al*; the value for PAA/PEO also shows good agreement with the data of Papisov at 550 J mol^{-1}. Values in 50% methanol solution of −700 and −750 J mol, also support the data of Papisov.

Although the calorimetric data on which figure 46 is based showed no evidence for a 1:1 complex, it is instructive to determine the value of ΔH at a PMAA/PEO ratio of 1:1 for PEO values greater than 10000. The result is approximately 1100 J mol^{-1} at 10 °C, confirming the magnitude of the value of ΔH obtained by Papisov and Tsuchida.

Determination of the stability constant of the PMAA/PEO complex[118] as a function of temperature between 10 °C and 65 °C, yielded values of ΔH for complex formation which, although endothermic, were not in particularly good agreement with other data; the data do, however, show a marked variation with temperature for lower masses of PEO, again a trend shown in figure 46.

The general conclusion from the thermodynamic evidence of an endothermic spontaneous process of complexation is that positive values for the entropy of complexation must result – a classic illustration of the effects of hydrophobic interaction accompanied by a loss of order in the displaced water; the effects are more clearly seen in the case of PMAA rather than PAA with further reinforcement of the conclusion from the reversal in sign of ΔH in methanol/water systems.

Osada also considers the effect of ethanol concentration in ethanol/water systems upon complex formation – at temperatures up to 30 °C the degree of linkage achieves a maximum value at approximately 20 wt% ethanol concentration, decreasing slightly at 23 wt%, but collapsing at 37 wt% ethanol; this kind of behaviour is consistent with the maximum structuring effects of ethanol in water – the large increase in the degree of linkage, 0.08 to 0.46, when compared to water is further support for hydrophobic interactions.

Comparable studies of the energetics of complex formation between PMAA and PVP by Osada[118] and Tsuchida[117], both show larger endothermic ΔH values when compared with complex formation with PEO, 5880 J mol^{-1} as against 1300 J mol^{-1}; the data thus again indicate a larger hydrophobic interaction term when comparing the behaviour of PVP with PEO, whether in solution or in complex formation.

6. Sorption behaviour of water-soluble polymers

The adsorption of water-soluble polymers to surfaces in which the surface area to bulk volume ratio is very high, such as colloidal dispersions, can have profound effects upon the stability of that system in resisting coagulation. The molecular architecture of the polymer is of fundamental importance for the manner in which adsorption occurs, thus the various factors which have already been a subject for discussion are again reflected in the adsorption process.

Adsorption describes the molecular situation in which the concentration of a molecular species near an interface differs from that in the bulk solution, far from that interface; since the difference can be greater or smaller, adsorption is a relative process and therefore best described by the term 'surface excess concentration' – essentially the difference in concentration between the surface layer and an equilibrium layer in the bulk of the solution. If the surface area of the adsorbent is known then the surface excess concentration of the adsorbing species, the adsorbate, is expressed in the units of mg m^{-2}.

Adsorption is a function of the bulk concentration of the adsorbate, up to the limiting saturation condition, which, in turn, is a reflection of the number of adsorption sites, not necessarily the total surface coverage of the adsorbate – this situation is illustrated qualitatively in figure 48, for what are known as high, medium and low affinity systems. At low degrees of surface coverage adsorbed polymer molecules are likely to be isolated from each other, with no lateral interaction between them, those portions of the macromolecule in contact with the surface are known as 'trains', those extending into the solution being either 'loops' or 'tails'.

The minimum Gibbs energy situation of the adsorbed polymer molecule must reflect the balance of enthalpy and entropy changes occurring upon adsorption. The enthalpy change must, in turn, reflect the number of segment/solvent, solvent/surface and segment/surface contacts, whilst the entropy change reflects, in addition to changes of configurational entropy,

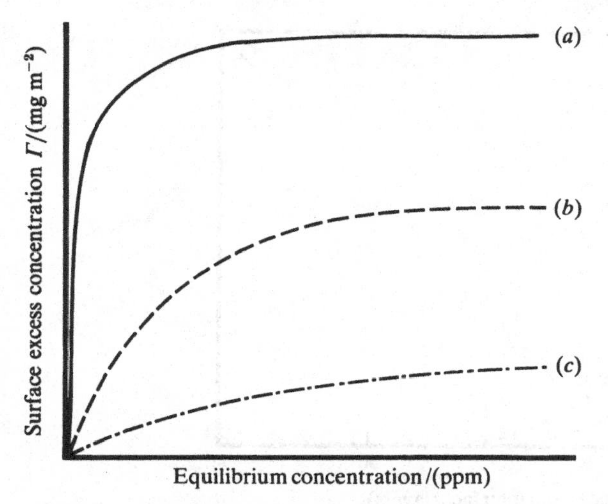

Figure 48. Form of the adsorption isotherm, plotted as function of equilibrium concentration, reflecting (*a*) high, (*b*) medium and (*c*) low affinity behaviour.

changes in the thermal entropy of the polymer and entropy changes due to the changed state of water molecules, either desorbed from the surface or released by the polymer.

The major consequence of the total interaction, excluding the special situation of a polycation adsorbing to a negatively charged surface, and vice versa, is that homopolymers are unlikely to adsorb strongly whilst copolymers, with hydrophilic and hydrophobic regions, have the capacity to adsorb strongly. Basically the situation is that a very hydrophilic polymer, one that dissolves easily in water, is unlikely to be a good adsorbate; poor aqueous solubility would, of course, lead to good adsorption – copolymers, with a balance of hydrophilic and hydrophobic groups therefore have the potential to satisfy these conflicting requirements.

The manner in which the macromolecule adsorbs to the surface is fundamental to its success as a 'stabiliser' of colloidal systems – a flat adsorbed conformation, with long trains, small loops and tails results in poor stability performance; a configuration with short trains, large loops and long tails is likely to be an effective stabiliser.

Several models have been proposed to describe the adsorption process, one of the more recent and comprehensive being the mean field approach of Scheutjens and Fleer[119], which, in turn, is based upon the Flory–Huggins lattice theory of polymer solutions, but with a mean field approximation of the possible configurations of polymer segments and solvent molecules in each lattice layer parallel to the surface and including a differential adsorption energy term for interactions between segments and surface. Predictions arise from the model of the extent of train, loop and tail

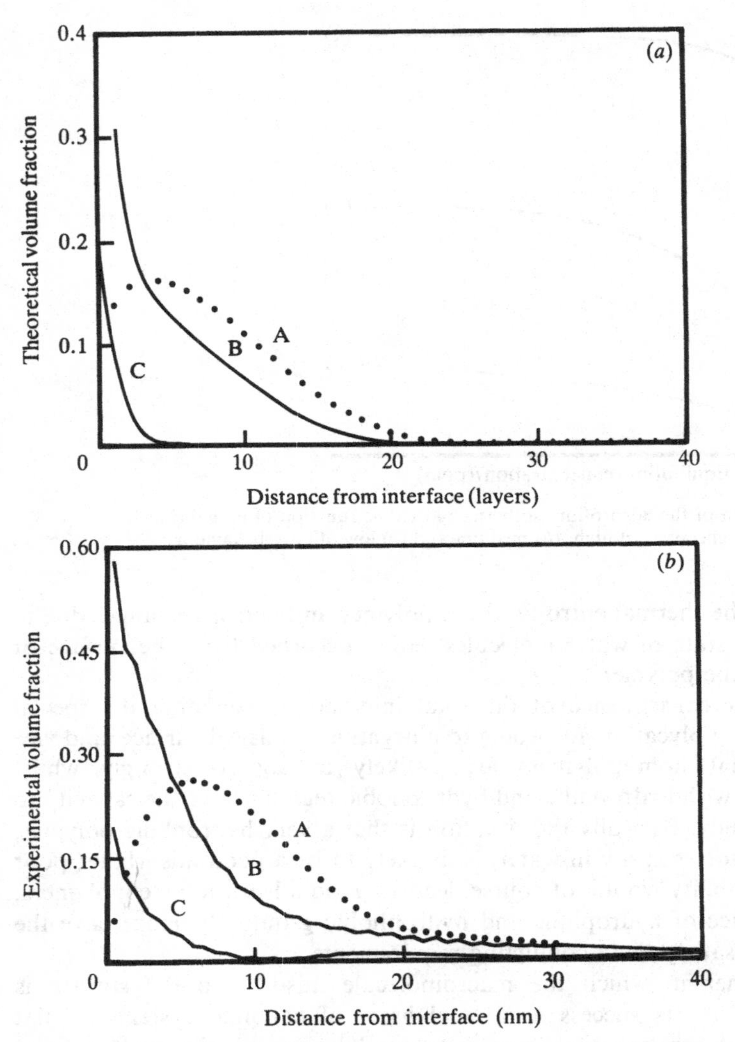

Figure 49. (*a*) Theoretical profiles for adsorbed polymers: dotted line, $\chi_s = 0.0$; full line, $\chi_s = 0.6$: A and B terminally anchored, C adsorbed from solution, chain length = 100 segments, $\chi_s = 0.4$. (*b*) Profiles of PEO adsorbed on polystyrene latex in water: dotted line, low surface charge density; full line, high surface charge density: A and B terminally anchored, C adsorbed from solution. Adsorbed amounts are 4 mg m^{-2} (A and B) and 0.7 mg m^{-2} (C); $\chi_s = 0.4$ (data from ref. 120).

formation, together with an estimate of the root-mean-square (rms) thickness of the adsorbed polymer layer; the predictions suggest that the aspect of greatest significance, with regard to adsorption is the length of the tails, rather than the size of the loops and trains. From these considerations the model predicts that the rms film thickness is proportional to $M^{\frac{1}{2}}$.

Cosgrove, Heath, Ryan and van Lent[120] have recently suggested a modification of the Scheutjens and Fleer model to consider the case of terminally attached chains, i.e. all chains that do not have their first segment in the first layer. Figure 49(*a*) shows the theoretical volume fraction profile for a polymer on a hexagonal lattice, A and B corresponding to a nett adsorption energy $\chi_s = 0$ and 0.6; for $\chi_s > \chi_c$, the critical adsorption energy (0.29 for a hexagonal lattice), the profile falls monotonically with distance, but for $\chi_s < \chi_c$ the chain is repelled and the profile shows a pronounced maximum. Curve C illustrates the situation for pure adsorption from solution for a χ_s value of 0.6. Figure 49(*b*) illustrates experimental data obtained by small-angle neutron scattering (SANS) of PEO adsorbed onto a low and high negative surface charge polystyrene surface (A and B respectively), together with curve C for adsorption from solution. Although, as the authors admit, it is difficult to scale the data to the lattice model exactly, the qualitative agreement with the theoretical predictions is very satisfactory; the high density of 'tails', particularly in the case of the terminally attached polymer is shown by the extent of the profile, well beyond the radius of gyration of the free chains in solution (3 nm) and almost equal to the extended length (39 nm). It should be noted that although the surface of the polystyrene latex is regarded as being notably hydrophobic, the greatest adsorption observed in this case is associated with the surface carrying the high quantity of charge – this being due to the presence of charged sulphate groups, arising from the use of a persulphate initiator in the polymerisation process. Cosgrove, Heath, Ryan and Crowley[121] have also reported SANS measurements of adsorption of a PVOH/VA copolymer (12 % residual VA) on a polystyrene latex surface, which also shows the existence of an appreciable 'tail' component in the segment distribution away from the surface; unfortunately in this work the authors do not give any details with regard to the characterisation of the polystyrene surface and the sample of PVOH/VA used is atypical of 'normal' PVOH/VA copolymer behaviour, due to the presence of a significant degree of chain branching in the molecule.

Croxton has also considered[122, 123] the development of boundary configurations for a terminally attached polymer molecule, using a self-avoiding hard sphere sequence, rather than the lattice model – in this case also the predictions indicate a majority contribution from 'tails'. In a later contribution[124] Croxton also included a direct assessment of the solvent contribution to the model, which had the effect of increasing the contribution of loops at the expense of trains, but with the latter still being very much the dominant components. For a sequence of 14 spheres, loops accounted for 28 %, as against 68 % for tails, the remaining 4 % being in trains.

The adsorption isotherms of PVOH/VA copolymers, as a function of molar mass, on a polystyrene latex surface show good agreement with the $M^{\frac{1}{2}}$ relationship predicted by Scheutjens and Fleer[119], a similar relationship

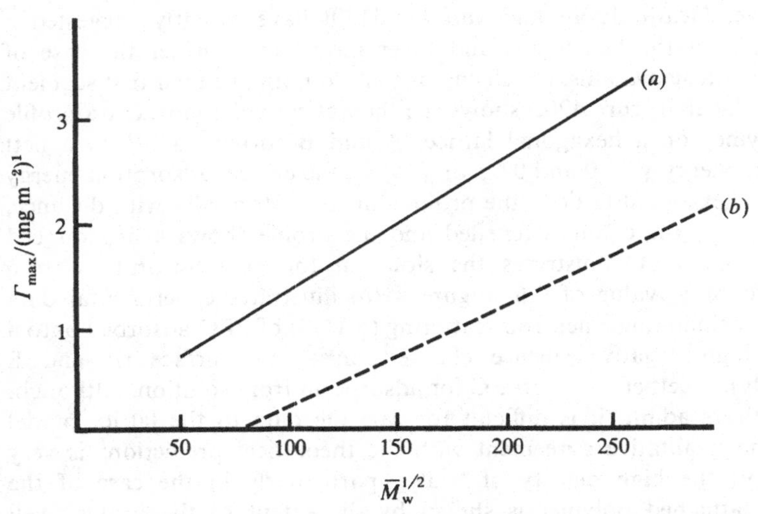

Figure 50. Equilibrium adsorption maxima versus $(\bar{M})_w^{\frac{1}{2}}$, for adsorption of a PVOH/VA copolymer onto (*a*) polystyrene and (*b*) PVA surfaces (data from refs 154 and 155).

being observed for adsorption onto a PVA latex surface, as shown in figure 50, but the degree of surface coverage being much lower in the latter case. Unfortunately it is not possible to isolate the determining factors for the differences since the nature of the adsorbing surfaces was insufficiently characterised. It should, however, be noted that persulphate was used as initiator in both cases but the VA surface is likely to contain an appreciable proportion of COO^- groups (from hydrolysis of the acetate groups). The significance of this kind of information is illustrated by figure 51, showing the effect of surface charge, due to sulphate groups arising from the persulphate initiator, on the extent of adsorption of a PVOH/VA copolymer (12% residual VA) on polystyrene lattices of closely similar particle size[125]. Adsorption clearly is a function of surface charge, yet for this classical adsorption situation it is generally accepted[126] that hydrophobic interaction is the major contributor. The data of Cosgrove *et al.* on the adsorption of PEO as a function of surface charge (SO_4^{2-}) on polystyrene, in conjunction with these data, suggest that the adsorption process is more complicated in character than at first appears – adsorption may well be by hydrophobic interaction but the presence of the sulphate groups on the particle surface appears to have a similar 'quality of solvent' effect upon solvent layers adjacent to the particle surface, as does the ion in bulk water, making this water 'worse' for hydrophobic regions, thus promoting adsorption.

Adsorption of PEO onto a highly charged polystyrene surface is clearly of 'high affinity' character and increases with increasing molar mass of the

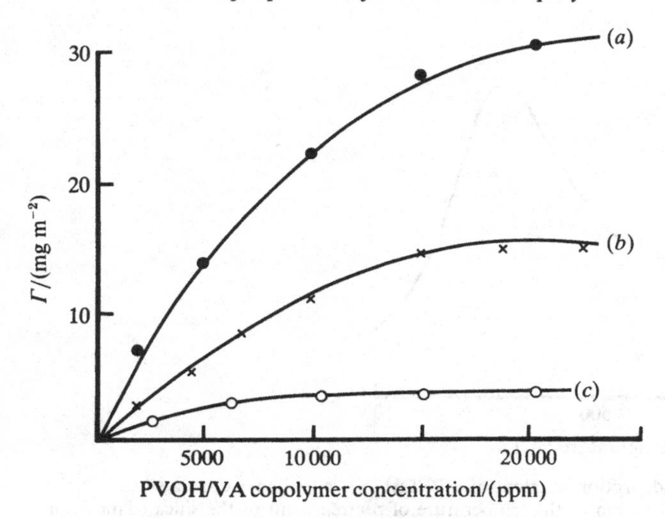

Figure 51. Adsorption isotherms of a PVOH/VA copolymer (12% residual VA) on polystyrene latex surface as a function of surface charge density: (*a*) 6.6×10^{-5} C cm^{-2}; (*b*) 1.9×10^{-5} C cm^{-2}; (*c*) 0.61×10^{-5} C/cm^{-2} (data from ref. 125).

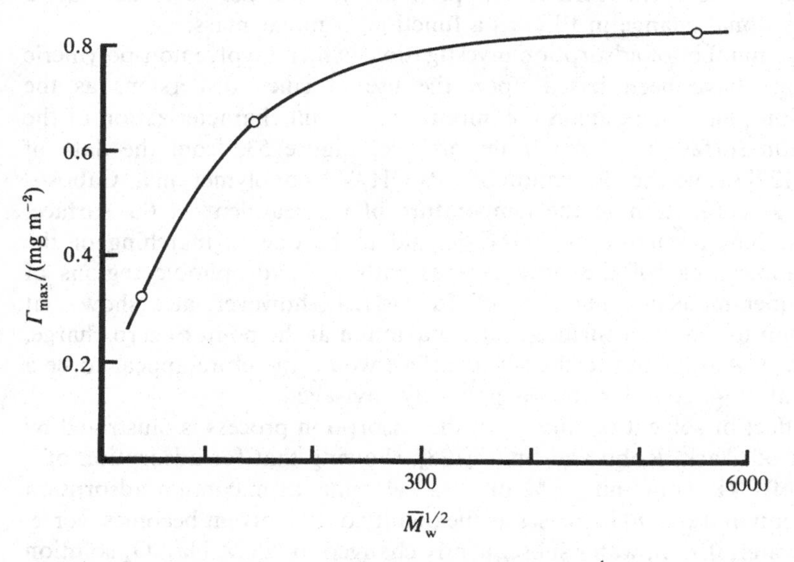

Figure 52. Equilibrium adsorption maxima of PEO versus ($M_w^{\frac{1}{2}}$) onto a polystyrene latex surface (data anon).

polymer, but as the data illustrated in figure 52 shows, adsorption is not a linear function of $M_w^{\frac{1}{2}}$, as predicted by the model of Scheutjens and Fleer; the data, unfortunately, are limited, but the change of slope does appear to occur at around the same molar mass as the discontinuity observed for PEO when

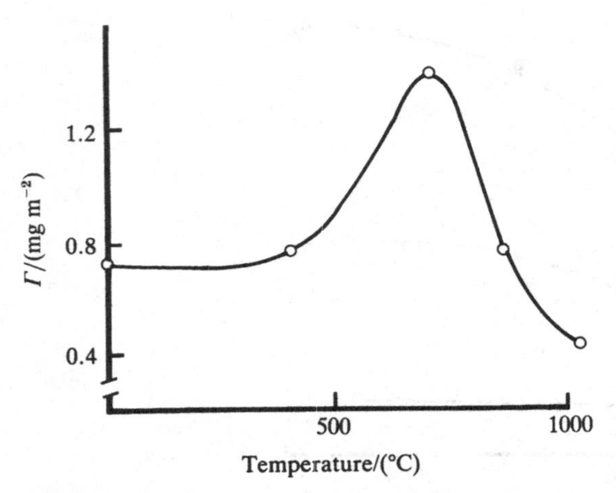

Figure 53. Equilibrium adsorption maxima of a PVOH/VA copolymer onto a silica dispersion, plotted as a function of the temperature of pretreatment of the silica (data from ref. 127).

involved in complex formation, perhaps yet further evidence of a conformational change in PEO as a function of molar mass.

A large number of adsorption investigations which involve non-polymeric adsorbents have been based upon the use of silica dispersions as the adsorption phase; here again the importance of full characterization of the adsorption surface is of major importance. Figure 53, from the data of Tadros[127] shows the adsorption of a PVOH/VA copolymer on a 'Cabosil' surface, as a function of the temperature of pretreatment of the surface, the maximum occurring at 700 °C is said to be due to matching of the hydrophobic areas of the silica surface with the hydrophobic regions of the polymer molecule. The data of Tadros[127], however, also show that adsorption to the silica surface is at a maximum at the point of zero charge, pH of 2.65. Adsorption to the silica surface would therefore appear to be a great deal more complex than is presently envisaged.

The effect of solvent quality upon the adsorption process is illustrated by the data of Clark, Robb and Smith[128], showing that for adsorption of a sample of PVP, containing 3% of an alkylamine, at maximum adsorption the fraction in trains(p) increases as the quality of the solvent becomes worse, 0.57 in water, 0.75 in water subsequently changed to 0.5 M Na_2SO_4 solution and 0.54 in 0.5 M Na_2SO_4 solution subsequently changed to water – 0.1 M NaCl solution was without effect upon 'p'. The effect of making the solvent 'worse' is to flatten the conformation of the adsorbed polymer molecule, increasing the proportion of trains at the expense of loops and tails.

Estimates of the thickness of the adsorbed polymer layer have frequently been based on hydrodynamic measurements[125]; it is not yet clear how this 'effective' hydrodynamic thickness relates to the equilibrium adsorbed thickness, which can, in principle, be obtained from SANS measurements. An alternative technique for determination of adsorbed polymer layer thickness is the direct measurement of the force of interaction between two smooth solid surfaces, usually mica or quartz, bearing the adsorbed polymer layer and immersed in the solvent medium. Klein and Luckham[129] have reported such studies of adsorbed PEO in a 'good' solvent, 0.1 M KNO_3 solution, the extension from the surface of the tails, often taken as the Flory radius[130, 131] i.e. the swollen end-to-end distance, should lead to interaction when the two surfaces are separated by twice the Flory distance; in fact these workers report that interaction occurs at less than two thirds of this distance, implying that considerable interpenetration of tails occurs before a repulsive interaction can be detected.

Goetze, Sonntag and Rabinovitch[132] have also recently reported direct force of interaction measurements, using quartz fibres coated with PVP and PVOH samples of varying molar mass; utilising the Hesselink, Vrij and Overbeek theory[133] of adsorbed chain interaction the authors report that, in the case of lower molar mass materials, the interaction can be accounted for by interaction of tails containing relatively small numbers of segments – no satisfactory agreement is however, obtained with higher molar mass materials. The conclusions are at variance with expectation and the authors do not attempt to correlate the results with data derived from hydrodynamic measurements; the authors do, however, show that, for similar molar masses, the adsorbed layer of PVOH/VA copolymer containing 2% residual VA is much more expanded than that of one containing 12% residual VA. Rather surprisingly the authors report that a 'homopolymer' of PVOH, although of lower molar mass does still exhibit an appreciable energy of interaction; this is unexpected since the homopolymer shows little stabilising capability for colloidal dispersions.

Goetze and Sonntag[134] have also recently investigated the role of solvent quality in the interaction of two layers of adsorbed PVOH, using the same technique. The thickness of the adsorbed polymer layer was found to be equivalent to several radii of gyration of the free coil in solution and a lyotropic series was observed in the thickness of the layer in the presence of alkali metal cations, increasing in the sequence from Li^+ to Rb^+. This sequence is in reverse order when considering the quality of solvent effect of these ions, the Li^+ ion, as a 'salter-in' might be expected to produce a more extended conformation of the PVOH, with Rb^+ ion producing the least. The adsorption surfaces, however, were quartz fibres and silica, surfaces which are known to adsorb a layer of water molecules, which, in turn, influences the mechanism of polymer adsorption to the surface[135] – as a consequence a reversed Hofmeister series is observed, reflecting hydrophilic rather than hydrophobic effects.

Brown and Rymden[136] have examined the adsorption behaviour of semi-synthetic water-soluble polymers by observing the diffusion of poly-

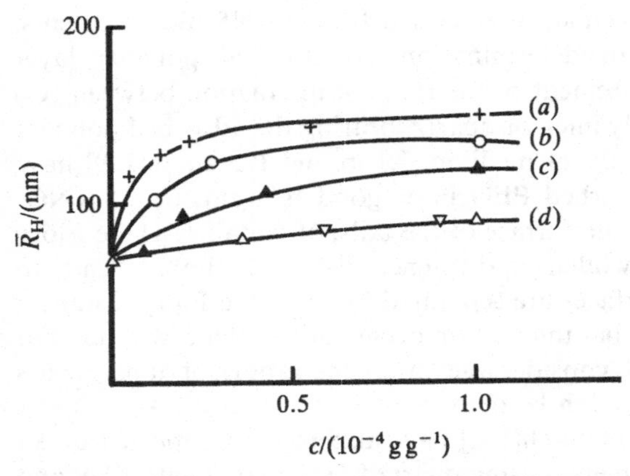

Figure 54. Hydrodynamic radii, derived from diffusion coefficients of a polystyrene latex in very dilute solutions of various polymers: (*a*) HEC; (*b*) HPC; (*c*) CMC, pH 9; (*d*) CMC, salt free (data from ref. 136).

styrene latex spheres in polymer solutions, using dynamic light-scattering techniques; figure 54 illustrates the apparent hydrodynamic radius R_H of the latex particle in solutions of HEC, HPC, CMC (at high and low pH) – adsorption is clearly greatest in HEC solution, becoming negligible in CMC at low pH. The magnitude of R_H is related to a bimodal analysis of the data, a fast component, corresponding to the radius of the latex sphere and a slow component whose value corresponds to R_H. The authors state that the hydrodynamic thickness of the adsorbed polymer layer is approximately twice the radius of gyration of the free polymer, thus if the radius of gyration of HEC is 71.5 nm and the plateau value of R_H is 150 nm, this implies some bridging between two particles by two HEC macromolecules. This situation is, of course, completely at variance with the measurements of Klein and Luckham and may be a consequence of relating the unrelatable, hydrodynamic layer thickness and 'actual' layer thickness, or at least thickness determined by direct measurement.

Such an explanation may account for the discrepancy reported by Cosgrove, Crowley and Vincent[137] between SANS and photocorrelation spectral studies (PCS) of latex/PEO – SANS results gave lower results for the increase in apparent radius in the presence of PEO than PCS; the SANS technique, however, is not sensitive to the small amount of polymer participating in bridging whereas PCS is very sensitive to such a change.

Very little information exists on the kinetics of the adsorption of water-soluble polymers to a solid substrate surface, rather than to the air/water interface[138]; the recent contribution by Pfefferkorn, Carroy and Varogui[139], who investigated the adsorption of radioactively tagged PAAm on a modified silica surface, is therefore particularly interesting. Modification of the silica bead surface produced surfaces with varying aluminol (AlOH)

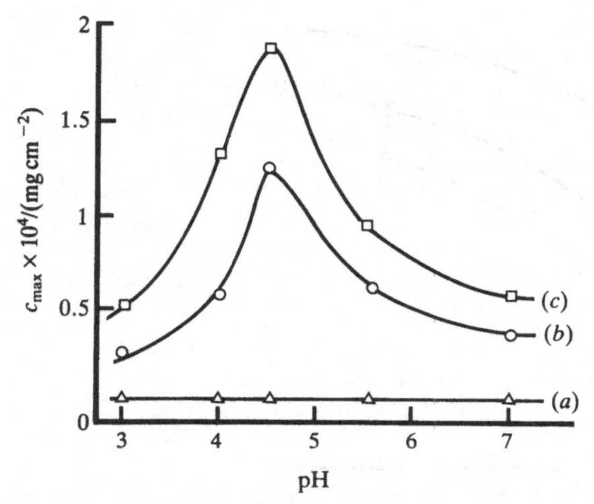

Figure 55. Maximum adsorption of PAAm as a function of pH and [AlOH]/[SiOH] ratio: (*a*) nonmodified beads; (*b*) 5% aluminol grafting; (*c*) 12% aluminol grafting. PMAA $\bar{M}_w = 1.2 \times 10^6$ (data from ref. 139).

contents on which the PAAm showed high affinity adsorption characteristics, which followed a pH dependence, as shown in figure 55. No detectable adsorption occurred on the unmodified silica surface and the maximum observed as a function of pH on the modified surfaces is due to the aluminols being in the $AlOH_2^+$ state below pH 4.5, whilst above 4.5 the AlO^- form is present. The kinetics of adsorption revealed two domains; initially adsorption increased rapidly until 60–70% of available sites were filled, after which adsorption showed a marked decrease in rate to approach the asymptotic value very slowly. The difference between the two rates shows a strong temperature dependence, *viz*: 15 °C, 3.5; 25 °C, 8.6; 35 °C, 8.8; 45 °C, 2.4; 55 °C, 1.7; which the authors suggest may be due to the adsorbed molecule undergoing a conformational change at higher surface coverage. It is the case that, with the exception of the two highest temperatures, the amount adsorbed at 60–70% of surface coverage is considerably in excess of that to be expected if the adsorbed macromolecule conformation is similar to that of the molecule in solution (free surface should be zero at $\phi = 0.1$ for a pH of 4.5) – such a change suggests the likelihood of a considerable change in conformation of the adsorbed species; curiously the authors do not make any comment upon the relatively low value of the ratio at 15 °C, which coincides with the maximum amount of adsorption.

Adsorption data determined on clay surfaces may appear as somewhat unusual, but a major use of high molar mass water-soluble polymers is in enhanced oil recovery. Figure 56[140] illustrates the adsorption of CMC onto peptised sodium montmorillonite as a function of the DS of the CMC and concentration of sodium chloride[141]; with increasing salinity CMC adsorbs

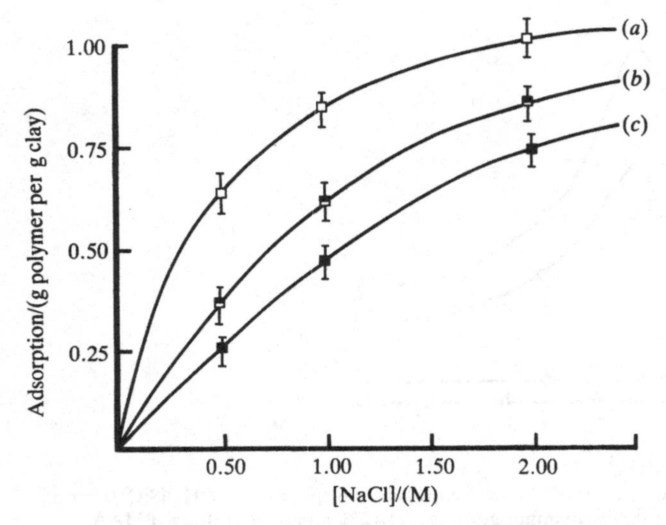

Figure 56. Adsorption (g/g) dependence upon the DS of CMC: (*a*) 0.99; (*b*) 1.19; (*c*) 1.46, and upon the sodium chloride concentration in the aqueous phase. Substrate is peptised sodium montmorillonite (date from ref. 140).

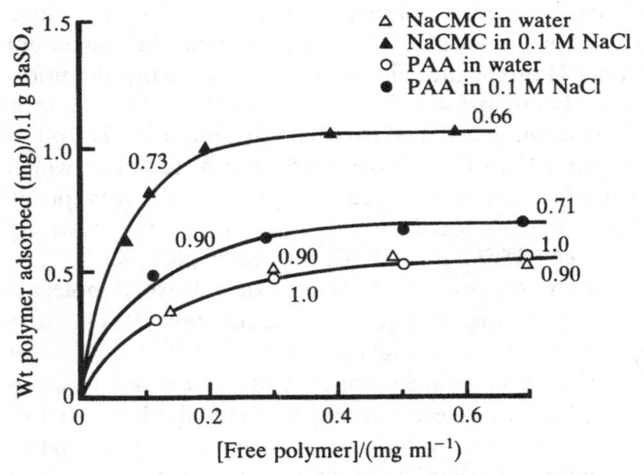

Figure 57. Adsorption isotherms of NaCMC ($\bar{M}_v = 1 \times 10^5$) and PAA ($\bar{M}_v = 8 \times 10^4$), onto barium sulphate at pH 9 in water and in 0.1 M NaCl. Numbers correspond to fraction of segments in trains (data from ref. 141).

in inverse relationship to its degree of substitution. A combination of two effects seems apparent here, the higher salinity values must affect the quality of the solvent and increase the flexibility of the macromolecule.

Figure 57 illustrates the adsorption of NaCMC and PAA onto barium sulphate (used as a weight control agent in drilling muds) at high pH (= 9)

[142]; adsorption of PAA increases only slightly on addition of sodium chloride (0.1 M), but that of NaCMC increases markedly and is associated with a large decrease in the fraction of the polymer existing as trains – most probably the consequence of exceeding ξ_c with a concomitant change in flexibility.

Glass, Ahmed and Karunasena[142] have investigated the adsorption of PEO as a function of concentration and molar mass onto different peptised cationic montmorillonites; a time dependence in adsorption onto ammonium montmorillonite was observed, but not in the case of the Ca^{2+} clay, adsorption in the first case being accompanied by greater symmetry and increased ordering of the clay as a function of PEO molar mass and concentration. Studies of competitive adsorption onto the ammonium clay showed that PEO of any molar mass would displace cellulose ethers (of any MS level); this preference may be due to a greater hydrophobic interaction by the PEO *vis-a-vis* the substituted cellulose ethers, or alternatively the PEO may be more successful in penetrating the layer-like structure of the ammonium clay – more extensive data covering the whole range of alkali metal salts would be very interesting in this respect.

One of the most important industrial uses of water-soluble polymers is as stabilisers for colloidal systems, allowing controlled flocculation of the system or enhanced stability, depending upon the degree of surface coverage. It is not possible to give a detailed explanation of the processes in this limited review, the reader is therefore referred to the excellent review of the subject by Napper[143]; suffice to say that an adsorbed polymer layer introduces an additional Gibbs energy term ΔG_s into the energetics of interaction between colloidal particles. If the term is positive then the additional factor is termed 'steric stabilisation' and since ΔG_s is equivalent to $\Delta H_s - T\Delta S_s$ and $d(\Delta G_s/dT) = -\Delta S_s$, where ΔG_s, ΔH_s and ΔS_s are the Gibbs energy, enthalpy and entropy of steric stabilisation respectively, three possible pathways to steric stability are possible:

(i) ΔH_s negative, ΔS_s positive, with $\Delta H_s/\Delta S_s \ll 1$, termed entropic stabilisation, thus flocculation occurs on cooling.

(ii) ΔH_s positive, ΔS_s positive, with $\Delta H_s/\Delta S_s \gg 1$, termed enthalpic stabilisation, flocculation occurs on heating.

(iii) ΔH_s positive, ΔS_s negative, with $\Delta H_s <> \Delta S_s$, a combination situation, usually very stable against flocculation.

Comparison of the three types with the solubility behaviour of water-soluble polymers shows that close similarity exists between solubility and flocculation behaviour, thus type (i) typifies nonaqueous systems whilst type (ii), enthalpic stabilisation, is more typical of aqueous systems.

As in the case of the polymers in solution, a critical value of χ (χ_c) arises, at which flocculation will occur for the adsorbed polymer layer; all the factors that have been considered in determining the state of solution of water-soluble polymers will therefore play similar roles in the stability of a

Table 10. *Theta and critical flocculation temperatures*

Polymer	M.Wt.	Dispersion medium	CFPT/(K)	U/L	θ/(K)
PEO	10000	0.39 M $MgSO_4$	318	U	315
	96000		316		315
	1000000		317		315
PAA	9800	0.2 M HCl	287	L	287
	51900		283		287
	89700		281		287
PVOH	26000	2.0 M HCl	302	U	300
	57000		301		300
	270000		312		300
PAAm	18000	2.1 M $(NH_4)_2SO_4$	292	L	—
	60000		295		—
	180000		280		—

sterically stabilized system. However, since the adsorbed polymer has been shown, in general, to have a different conformation in the adsorbed state when compared to that in solution, critical values, such as LCST, are not going to be the same as those of the free polymer; Napper[143] has reported the influence of molar mass upon the critical flocculation temperatures of several adsorbed copolymers, shown in table 10. PEO and PVOH exhibit upper critical flocculation temperatures (UCFT) whereas PAA and PAAM show lower critical flocculation temperature (LCFT) behaviour; the data, however, are obtained in a variety of salt solutions, which must raise a question mark over their general applicability, particularly those of PEO and PAAm where that well-known 'salter-out' the sulphate ion is present! Furusawa and Kimura[144] have determined the adsorption of HEC and PVOH on polystyrene latex at the LCST of the polymer and its effect upon colloid stability; adsorption at the LCST was found to be 1.5 times larger than that determined at room temperature – the more dense adsorbed layer showing greater protective action against both the effects of salt and temperature, the greater protective action remaining for a period of several days.

Recently Okuba[145] investigated the stabilising capabilities of a range of water-soluble polymers by using reflectance spectrum measurements to observe the face-centred lattice structure of high volume fraction monodisperse polystyrene lattices and the effect of the polymers in solution upon this structure; the following sequence of intersphere distances within the ordered phase was observed:

PEO > PVOH > no polymer > PVP = HEC > PAAm

The sequence confirms the data reported earlier, that PEO will displace substituted cellulosic ethers, implying that in a competitive situation these polymers should preferentially displace each other from the surface. Such behaviour has considerable relevance to the industrial uses of water-soluble polymers and the recent paper of Csempesz, Rohrsetzer and Korrass[146] is thus especially interesting. These workers studied the adsorption of MC, PVOH and PVP, and their binary mixtures on an As_2S_3 sol and a polystyrene latex, and their effects upon the stability of the dispersions. Each polymer exhibited a high affinity isotherm for both types of surface, but in adsorption from binary mixtures the order of preferential adsorption was PVP > MC > PVOH on the As_2S_3 sol and MC > PVP > PVOH on the polystyrene. Stabilisation efficiency correlated with the order of preferential adsorption but, in addition, synergism was exhibited in the case of MC/PVP mixtures on the As_2S_3 sol, which was explained by the authors as due to a large difference in the segment differential adsorption energies of the competing polymers for the surfaces and the more weakly adsorbing polymer being a more effective stabiliser than the preferentially adsorbing species. These data clearly show the importance of knowledge of the adsorbing surface for understanding the adsorption process; in addition this particular example is an excellent illustration of the various factors discussed in this review, in particular, detailed knowledge of the molecular architecture of the polymer. In conclusion, as a final illustration of the importance of solvent quality, Rohrsetzer and Csempesz[146] reported that the presence of potassium nitrate increased sol stabilisation, PVP showing a lower stabilising layer thickness than PVOH or MC for equally adsorbed amounts, clearly reflecting the 'better' conditions pertaining in the solvent due to the presence of the 'salting in' ion, NO_3^-, and how these conditions influence an individual polymer.

7. Conclusion

The solution behaviour of water-soluble polymers is akin in many ways to Salome's 'Dance of the Seven Veils', the removal of each veil makes the problems even more tantalising – if this review helps to lift even the corner of one veil, then the author will be well satisfied; no one, however, should be in any doubt – we are still a long way from removing the seventh veil!

Appendix: Glossary of water-soluble polymers

Acrylic Polymers		
	PAA	Polyacrylic acid
	PAAm	Polyacrylamide
	PDMAAm	Poly-(N,N-dimethylacrylamide)
	PIPAAm	Poly-(n-isopropylacrylamide)
	PMAA	Polymethacrylic acid
	PMAAm	Polymethacrylamide
	PCA	Polycrotonic acid
	PMA	Polymaleic acid
	PEA	Polyethacrylic acid
Vinyl Polymers		
	PESA	Polyethylenesulphonic acid
	PSSA	Polystyrenesulphonic acid
	PVOH	Polyvinylalcohol
	PVAm	Polyvinylamine
	PVMA	Polyvinylmethoxyacetal
	PVME	Polyvinylmethylether
	PVEE	Polyvinylethylether
	PVBE	Polyvinylbutylether
	PVHE	Polyvinylhexylether
	PVMO	Polyvinylmethyloxazolidone
	PVP	Polyvinylpyrrolidone
	P4VP	Poly-4-vinylpyridine
	P4VPO	Poly-4-vinylpyridine-N-oxide
	PVSA	Polyvinylsulphuric acid
Polyoxides and imines		
	PEI	Polyethyleneimine
	PEO	Polyethyleneoxide
(alternatively	PEG	Polyethyleneglycol)
	PPO	Polypropyleneoxide
(alternatively	PPG	Polypropyleneglycol)
	PMO	Polymethyleneoxide
Copolymers		
	EO/PO	Ethyleneoxide/propyleneoxide
	MA/VAE	Maleic acid/vinyl alkyl ether
	MA/AA	Maleic acid/acrylic acid
	MAAm/MAA	Methacrylamide/methacrylic acid
	VA/VOH	Vinylacetate/vinylalcohol

VA/VP	Vinylacetate/vinylpyrrolidone
VA/VSA	Vinylacetate/vinylsulphuric acid
MAA/MMA	Methacrylic acid/methylmethacrylate
AA/Am	Acrylic acid/acrylamide
S/VP	Styrene/vinylpyrrolidone
S/SSA	Styrene/styrenesulphonic acid

Semi-synthetics

Simple ethers of cellulose

MC	Methylcellulose
EC	Ethylcellulose
HEC	Hydroxyethylcellulose
HPC	Hydroxypropylcellulose
CMC	Carboxymethylcellulose

Mixed ethers of cellulose

EHEC	Ethylhydroxyethylcellulose
EMC	Ethylmethylcellulose
HPMC	Hydroxypropylmethylcellulose
HEMC	Hydroxyethylmethylcellulose
HBMC	Hydroxybutylmethylcellulose

Starch derivatives

SAC	Starchacetate
HES	Hydroxyethylstarch
CMS	Carboxymethylstarch
AAS	Aminoalkylstarch

References

1. F. F. Vercauteren, W. A. B. Donners, R. Smith, N. J. Crowther & D. Eagland, *Eur. Polym. J.* **23**, 711 (1987).

2. I. Croon, *Sven Papperstin* **63**, 247 (1960).

3. E. Wolfram & M. Nagy, *Koll. Z. Z. Polym.* **227**, 86 (1968).

4. D. Eagland, F. F. Vercauteren, Th. G. Scholte, W. A. B. Donners, M. D. Lechner & R. Mattern, *Eur. Polym. J.* **22**, 351 (1986).

5. P. J. Flory, '*Principles of Polymer Chemistry*', Cornell Univ. Press, New York, 1958.

6. M. L. Miller, K. O'Donnell & J. Skogman, *J. Colloid Sci.* **17**, 649, (1962).

7. T. Shiomi, K. Imai, C. Watanabe & M. Miya, *J. Polym. Sci.* **22**, 1305, (1984).

8. I. Sakurada, Y. Sakaguchi & Y. Itoh, *Kobunshi Kagaku* **14**, 41 (1957).

9. Y. Sakaguchi & Y. Itoh, *Kobunshi Kagaku* **15**, 635 (1958).

10. M. Ataman & E. A. Boucher, *J. Polym. Sci., Polym. Phys. Edn.* **20**, 1585 (1982).

11. P. K. Nandi & D. R. Robinson, *J. Amer. Chem. Soc.* **94**, 1308 (1972).

12. M. Garvey & I. D. Robb, *J. Chem. Soc. Faraday Trans.* **75**, 993 (1979).
13. H. G. Elias, *Light Scattering from Dilute Polymer Solutions*, (ed. M. B. Huglin) Academic Press, New York 1972.
14. L. Mrkvickova, E. Prokopova & O. Quadrat. *Colloid & Polym. Sci.* **265**, 987 (1987).
15. N. J. Crowther & D. Eagland, *J. Chem. Soc. Faraday Trans. I* **84**, (1988), in press.
16. D. Eagland & N. J. Crowther, *Faraday Symp. Chem. Soc.* **17**, 161 (1982).
17. F. Franks & T. Wakabayshi, *Z. Phys. Chem.* **155**, 171 (1987).
18. W. M. Kulicke, R. Kniewski & J. Klein, *Prog. Polym. Sci.* **8**, 373 (1982).
19. P. G. deGennes, *Macromol.* **9**, 587 (1976).
20. W. F. Polik & W. Burchard, *Macromol.* **16**, 978 (1983).
21. M. L. Fishman, L. Pepper & P. E. Pfeffer, *Water-Soluble Polymers*, (ed. J. E. Glass) Adv. Chem. Series, 213, ACS, Washington DC, 1986.
22. Z. I. Kertez, *The Pectic Substances*, Interscience, New York, 1951.
23. H. S. Owens, H. Lotzkar, T. H. Schultz & W. D. Maclay, *J. Amer. Chem. Soc.* **68**, 1628 (1946).
24. D. T. Pals & J. J. Hermans, *Recl. Trav. Chem. Pays-Bas* **71**, 458 (1952).
25. R. C. Jordan & D. A. Brant, *Biopolymers* **17**, 2885 (1978).
26. M. L. Fishman, P. E. Pfeffer, R. A. Barford & W. L. Doner, *J. Agri. Food Chem.* **32**, 372 (1984).
27. M. A. F. Davis, M. J. Gidley, E. R. Norris, D. A. Powell & D. A. Rees, *Int. J. Biol. Macromol.* **2**, 330 (1980).
28. V. D. Sorochan, A. K. Dzizinko, N. S. Boden & Y. S. Ovodo, *Carbohydr. Res.* **20**, 243 (1971).
29. D. E. Hanke & D. H. Northcote, *Biopolymers* **14**, 1 (1975).
30. M. L. Fishman, L. Pepper, P. E. Pfeffer & R. A. Barford, Abstracts of papers, 186th National meeting, Amer. Chem. Soc., Washington, DC, 1983.
31. F. M. Winnick, M. A. Winnick, S. Tazuke & C. K. Ober, *Macromol.* **20**, 38 (1987).
32. F. M. Winnick, *Macromol.* **20**, 2745 (1987).
33. K. Char, C. W. Frank, A. P. Gast & W. T. Tana, *Macromol.* **20**, 1823 (1987).
34. G. N. Malcolm & J. S. Rowlinson, *Trans. Faraday Soc.* **53**, 921 (1957).
35. P. Molyneux, in *Water, A Comprehensive Treatise*, Vol. 4, (ed. F. Franks) Plenum Press, New York, 1975.
36. C. J. Butler & D. Eagland, to be published
37. A. Kagamoto, S. Murakami & R. Fujishiro, *Makromol. Chem.* **105**, 154 (1967).
38. D. Eagland, G. C. Wardlaw & I. Thorn, *Colloid Polym. Sci.* **256**, 1073 (1978).
39. A. C. Sau, *Proc. Amer. Chem. Soc., PMSE Divn.* **57**, 497 (1987).
40. J. Bock, D. B. Siano, P. L. Valint & S. J. Pace, *Proc. Amer. Chem. Soc., PMSE Divn.* **57**, 487 (1987).
41. R. A. Gelman & H. G. Barth, in *Water-Soluble Polymers*, (ed. J. E. Glass) Adv. Chem. Series, 213, ACS, Washington, DC, 1986.
42. J. Goldfarb & S. Rodriguez, *Makromol. Chem.* **116**, 96 (1968).
43. W. W. Greassley, *Adv. Polym. Sci.* **16**, 1 (1974).
44. J. Schurz, *Rheol. Acta.* **14**, 293 (1975).
45. B. R. Breslau & I. F. Miller, *J. Phys. Chem.* **74**, 1056 (1970).

46. G. C. Wardlaw, S. Humphrey & D. Eagland, *Br. Polym. J.* **9**, 278 (1977).
47. A. Silberberg, J. Eliassaf & A. Katchalsky, *J. Polym. Sci.* **23**, 259 (1957).
48. G. S. Manning, *J. Phys. Chem.* **85**, 870 (1981).
49. N. Yoshida, *J. Phys. Chem.* **69**, 4867 (1978).
50. M. Satoh, T. Kawashima, J. Komiyama & T. Iijima, *Polym. J.* **19**, 1191 (1987).
51. M. Satoh & J. Komiyama, *Polym. J.* **19**, 1201 (1987).
52. C. T. Henningson, D. Karluk & P. Ander, *Macromol.* **20**, 1291 (1987).
53. P. Ander & W. Lubas, *Macromol.* **14**, 1058 (1981).
54. W. Lubas & P. Ander, *Macromol.* **13**, 318 (1980).
55. P. Ander & M. Kardon, *Macromol.* **17**, 2431 (1984).
56. P. Ander & M. Kardon, *Macromol.* **17**, 2436 (1984).
57. J. W. Klein & B. R. Ware, *J. Chem. Phys.* **80**, 1334 (1984).
58. H. Grasdalen & B. J. Kvam, *Macromol.* **19**, 1913 (1986).
59. M. Nagasawa, *Pure Appl. Chem.* **26**, 519 (1971).
60. M. B. Mathews, *Biochim. Biophys. Acta* **35**, 9 (1959).
61. M. Mandel, J. C. Leyte & M. G. Stadhouder, *J. Phys. Chem.* **71**, 603 (1967).
62. P. L. Dubin & U. P. Strauss, *J. Phys. Chem.* **74**, 2842 (1970).
63. V. Crescenzi, F. Quadrifoglio & F. Delben, *J. Polym. Sci. Pt A2* **10**, 357 (1972).
64. P. J. Martin & U. P. Strauss, *Biophys. Chem.* **11**, 397 (1980).
65. P. L. Dubin & U. P. Strauss, in *Polyelectrolytes and their Applications* (eds. A. Rembaum & E. Selegny), D. Reidel, Dordrecht, Holland 1975.
66. U. P. Strauss & G. Vesnaver, *J. Phys. Chem.* **79**, 2426 (1975).
67. B. W. Barbieri & U. P. Strauss, *Macromol.* **18**, 411 (1985).
68. J.-L. Hsu & U. P. Strauss, *Proc. Amer. Chem. Soc. PMSE Divn.* **57**, 603 (1987).
69. H. Daoust & A. Lajoie, *Can. J. Chem.* **54**, 1853 (1976).
70. M. Morcellet, C. Loucheux & H. Daoust, *Macromol.* **15**, 894 (1982).
71. J. Morcellet-Sauvage, M. Morcellet & C. Loucheux, *Makromol. Chem.* **182**, 949 (1981).
72. J. R. C. van der Maarel, D. Lankhorst, J. de Bleijser & J. C. Leyte, *Macromol.* **20**, 2390 (1987).
73. J. Plestil, Yu. M. Ostanevich, Yu. V. Bezzabotnov & D. Hlavata, *Polymer* **27**, 1241 (1986).
74. I. Nagata & Y. Okamoto, *Macromol.* **16**, 749 (1983).
75. Y. Muroga & M. Nagasawa, *Polymer* J. **18**, 15 (1986).
76. Y. Muroga, I. Sakuragi, I. Noda & M. Nagasawa, *Macromol.* **17**, 1844 (1984).
77. G. V. Schulz, W. Wunderlich & R. Kirske, *Macromol. Chem.* **75**, 22 (1964).
78. A. Arnold & J. Th. G. Overbeek, *Rec. Trav. Chim.* **69**, 192 (1950).
79. M. Nagasawa, T. Murase & K. Kondo, *J. Phys. Chem.* **69**, 4005 (1965).
80. C. Chang, D. M. Muccio & T. St. Pierre, *Macromol.* **18**, 2154 (1985).
81. Y. Kawaguchi & M. Nagasawa, *J. Phys. Chem.* **73**, 4382 (1969).
82. T. Kitano, S. Kawaguchi, N. Anazawa & A. Minakata, *Macromol.* **20**, 1598 (1987).
83. T. Kitano, S. Kawaguchi, N. Anazawa & A. Minakata, *Macromol.* **20**, 2498 (1987).

84. H. P. Gregor & M. Fredrick, *J. Polym. Sci.* **23**, 451 (1957).
85. N. Muto, T. Kamatsu & T. Nakagema, *Bull. Chem. Soc. Japan* **46**, 2711 (1973).
86. A. Minakata, K. Matsuma, S. Sasaki & H. Ohnuma, *Macromol.* **13**, 1549 (1980).
87. L. Zhang, T. Takernatsu & T. Norisuye, *Macromol.* **20**, 2882 (1987).
88. G. Paradossi & D. A. Brant, *Macromol.* **15**, 874 (1982).
89. T. Sato, T. Norisuye & H. Fujita, *Polym. J.* **16**, 341 (1984).
90. T. Sato, S. Kojima, T. Norisuye & H. Fujita, *Polym. J.* **16**, 423 (1984).
91. W. Liu, T. Sato, T. Norisuye & H. Fujita, *Carbohydr. R.* **160**, 267 (1987).
92. T. Coviello, K. Kajiwara, W. Burchard, M. Denline & V. Crescenzi, *Macromol.* **19**, 2826 (1986).
93. A. Jeanes, J. E. Pittsley & F. R. Senti, *J. Appl. Polym. Sci.* **5**, 519 (1961).
94. G. Holzwarth, *Biochemistry* **15**, 4333 (1976).
95. E. R. Morris, D. A. Rees, G. Young, M. D. Walkinshaw & A. Darke, *J. Mol. Biol.* **110**, 1 (1977).
96. J. G. Southwick, M. E. McDonnell, A. M. Jamieson & J. Blackwell, *Macromol.* **12**, 305 (1979).
97. S. Paoletti, A. Cesaro & F. Delben, *Carbohydr. Res.* **123**, 173 (1983).
98. I. T. Norton, D. M. Goodall, S. A. Frangou, E. R. Morris & D. A. Rees, *J. Mol. Biol.* **175**, 371, (1984).
99. M. Milas & M. Rinaudo, *Carbohydr. Res.* **158**, 191 (1986).
100. L. S. Hacche, G. E. Washington & D. A. Brant, *Macromol.* **20**, 2179 (1987).
101. E. Tsuchida, Y. Osada & K. Sanada, *J. Polym. Sci., Pt A1*, **10**, 2397 (1972).
102. E. Tsuchida & K. Abe, *Advances in Polymer Science*, **45**, Springer-Verlag, Heidelberg, 1982.
103. A. Nakajima, *Polym. J.* **7**, 550 (1975).
104. T. Ikawa, K. Abe, K. Honda & E. Tsuchida, *J. Polym. Sci.* **13**, 1505 (1975).
105. L. A. Bimendina, S. S. Saltybaeva & E. A. Bekturov, *Izv. Akad. Nauk. Kaz. SSR., Ser. Chim.* **6**, 22 (1978).
106. F. Franks & J. E. Desnoyers, *Water Science Reviews* **1**, 171 (1985).
107. E. A. Bekturov, L. A. Bimendina & G. S. T. Ileubaeva, *Vysokomol. Soedin, Ser. B.* **21**, 86 (1979).
108. A. D. Antipina, V. Yu. Baranorskii, I. M. Papisov & V. A. Kabanov, *Vysokomol. Soyed.* **A14**, 941 (1972).
109. Y. Osada & M. Sato, *Polym. Letters* **14**, 129 (1976).
110. B. Bednar, Z. Li. Y. Huang, L.-C. P. Chang & H. Morawetz, *Macromol.* **18**, 1829 (1985).
111. V. Yu. Baranovsky, A. A. Litmanovich, T. M. Papisov & V. A. Kabanov, *Eur. Polym. J.* **17**, 969 (1981).
112. H. T. Oyama, W. T. Tang & C. T. Frank, *Macromol.* **20**, 474 (1987).
113. H. T. Oyama, W. T. Tang & C. T. Frank, *Macromol.* **20**, 1839 (1987).
114. B. Bedman, H. Morewtez & J. A. Shafer, *Macromol.* **17**, 1634 (1984).
115. I. Iliopoulos & R. Audebert, *Eur. Polym. J.* **24**, 171 (1988).
116. I. M. Papisov, V. Yu. Baranorskii, Ye. I. Sergiera, A. D. Antipina & V. A. Kabanov, *Vysokomol. Soedin.* **A16**, 1113 (1974).
117. E. Tsuchida & K. Abe, *Adv. Polym. Sci.* **45**, 1 (1982).
118. Y. Osada, *J. Polym. Sci.* **17**, 3485 (1979).
119. J. M. H. M. Scheutjens & G. J. Fleer, *J. Phys. Chem.* **83**, 1619 (1979).

120. T. Cosgrove, T. G. Heath, K. Ryan & B. van Lent. *Polym. Comm.* **28**, 64 (1987).
121. T. Cosgrove, T. G. Heath, K. Ryan & T. L. Crowley, *Macromol.* **20**, 2879 (1987).
122. C. A. Croxton, *J. Phys. A:Math. Gen.* **19**, 987 (1986).
123. C. A. Croxton, *J. Phys. A:Math. Gen.*, **19**, 2353 (1986).
124. C. A. Croxton, *Polym. Comm.* **28**, 178 (1987).
125. D. Eagland, W. P. J. Baily & D. B. Farmer, to be published
126. Th. E. Tadros, *The Effect of Polymers on Dispersion Properties*, Academic Press, New York, 1982.
127. Th. F. Tadros, *J. Coll. Interface Sci.* **64**, 36 (1978).
128. A. T. Clark, I. D. Robb & R. Smith, *J. Chem. Soc. Faraday Trans.* I **72**, 1489 (1976).
129. J. Klein & P. L. Luckham. *Macromol.* **19**, 2007 (1986).
130. P-G. deGennes, *Macromol.* **14**, 1637 (1981).
131. P-G. deGennes, *Macromol.* **15**, 429 (1982).
132. Th. Goetze, S. H. Sonntag & Ya. Rabinovitch, *Coll. & Polym. Sci.* **265**, 134 (1987).
133. F. T. Hesselink, A. Vrij 6 J. G. Th. Overbeek, *J. Phys. Chem.* **75**, 2094 (1971).
134. G. T. Goetze & S. H. Sonntag, *Colloids Surf.* **25**, 77 (1987).
135. D. Eagland, in *Water, a Comprehensive Treatise*, Vol. 5 (ed. F. Franks), Plenum Press, New York, 1978.
136. W. Brown & R. Rymden, Macromol. **19**, 2942 (1986).
137. T. Cosgrove, T. L. Crowley & B. Vincent, *Adsorption from Solution*, Academic Press, New York, 1983.
138. D. E. Graham & M. C. Phillips, *J. Colloid Interface Sci.* **70**, 403 (1979).
139. E. Pfefferkorn, A. Carroy & R. Varogui, *Macromol.* **18**, 2252 (1985).
140. S. A. Heinle, S. Shah & J. E. Glass, *Water-Soluble Polymers* (ed. J. E. Glass), Adv. Chem. Series, **213**, ACS, Washington, DC 1986.
141. I. D. Robb & M. C. Cafe, *J. Colloid Interface Sci.* **65**, 82 (1982).
142. J. E. Glass, H. Ahmed, A. Karunasena, *Colloids Surf.* **21**, 335 (1986).
143. D. H. Napper, *Polymeric Stabilization of Colloid Dispersions*, Academic Press, New York, 1983.
144. K. T. Furusawa & K. Kimura, *Amer. Chem. Soc. Symp. Series* – Polymer Adsorption and Dispersion Stability, 186th Meeting, Washington, 1984.
145. T. Okuba, *J. Chem. Soc. Faraday Trans.* I, **83**, 2497 (1987).
146. F. Csempesz, S. Rohrsetzer & P. Korrass, *Colloids Surf.* **24**, 101 (1987).
147. S. Rohrsetzer & F. Csempesz, *Colloid Polym. Sci.* **264**, 992 (1986).
148. C. A. Finch, *Polyvinylalcohol, Properties and Applications*, Wiley, New York, 1973
149. H.-G. Elias, *Macromolecules, I – Structure and Properties*, Plenum, New York, 1984.
150. F. Franks, personal communication.
151. P. Molyneux, *Informal Discussion on Aqueous Solution Properties of Synthetic Polymers*, Cranfield, 1976.
152. D. Eagland & N. J. Crowther, to be published.
153. D. Eagland, to be published.
154. M. J. Garvey, Th. F. Tadros & B. Vincent, *J. Coll. Interface Sci.* **49**, 57 (1974)
155. W. P. J. Baily & D. B. Farmer, private communication.

Hydration of surfaces with particular attention to micron-sized particles

KENNETH E. NEWMAN

University of Sherbrooke, Sherbrooke, Quebec, J1K 2R1, Canada.†

1. Preamble

The interaction of solid materials with water is clearly of profound and vast scientific and technological importance. The range of areas where such understanding is important is so large as almost to defy listing. A highly incomplete but hopefully representative list might include mineral separation, electrodes, sensors and membranes, biocompatibility of non-biological material, geochemistry and weathering, oil recovery, archaeological dating, particle toxicology, composites and new materials, corrosion, cement and concrete, ice nucleation and meteorology, protein and biological assembly structure and stability, frost heaving, colloid stability, etc., etc.

In many ways study of the heterogeneous systems has mirrored that of homogeneous analogues; in particular we may note that solute–solute (homogeneous) and solute–surface (heterogeneous) studies have taken precedence over solute–solvent (homogeneous) and surface–solvent (heterogeneous) studies. The reasons for this are two-fold: firstly, relatively successful primitive electrostatic models for both homogeneous (Debye–Hückel) and heterogeneous (Gouy–Chapman–Stern) systems exist which do not require explicit treatment of the solvent beyond it being a dielectric continuum. Second, the study of both surface–solvent and solute–solvent interactions has been somewhat restricted by the limited range of experimental techniques available, although gas–surface study is very much a long-established research area. We mention in passing that relatively recently gas phase ion solvation studies have shed much light on the homogeneous systems.

The very large potential scope of this review implies that a certain number of topics of great relevance to the subject will not be discussed or will be given only passing mention. The very nature of water and of mineral particles particularly oxides, implies that surface hydrolysis and/or reaction is

† Present address: Chemistry Department, The King's College, 10766–97 Street, Edmonton, Alberta T5H 2M1, Canada

important and as such will be highly pH dependent. For other classes of solids, ion-exchange, surface dissolution and/or reaction will also be of great importance and will depend on electrolyte concentration. This whole area clearly impinges on colloid stability and/or electrode double layer studies but will only be discussed when directly relevant to hydration itself. Organic and biological surfaces will also not be discussed here, in spite of their undoubted importance, beyond a brief mention, in passing, to inorganic solids (most notably silica) with organic molecules chemically bonded to their surface.

Various aspects of the subject of this review were extensively treated in several chapters of Volume 5 of *Water, A Comprehensive Treatise*, published in 1978 under the editorship of Franks[1–3]. Thus, we will concentrate on the literature published since that time. The wealth of work published implies that it is almost impossible to produce an all-inclusive review. We thus plan to adopt a slightly different philosophy in this work in that we will discuss extensively the various experimental techniques available for the study of hydration of surfaces. We intend to underline the scope of information obtainable from each approach and also its limitations and shortcomings. The emphasis on varied experimental techniques arises from the fact that the interaction of water with solid surfaces is of such complexity that it is only by applying a battery of techniques in parallel to the same system, that we can make significant progress in our understanding; clearly, the larger the armoury we use, the more comprehensive will be our understanding. Within this discussion, we will include computer simulations of water/surface interactions. This area is still in its infancy but it already hints at the sorts of information that could become available in the future. Following this discussion of the experimental methods available, we will endeavour to give a 'thumbnail' sketch of the hydration properties of the various classes of inorganic materials. These will include both pure, synthetic solids as well as 'impure', geological materials.

2. Experimental methods

2.1. *Adsorption isotherms*

The measurement of adsorption isotherms has been a major part of the characterisation of micron-sized particles for very many years. Isotherms can be obtained in both the gas and liquid phase. However, in the latter case, it is the adsorption of a solute dissolved in a supposedly inert solvent that is studied. Variations of this approach might include preferential adsorption where two solutes compete for the adsorption sites or else studies in a binary solvent mixture (without solute) where the solvent components are in competition. This latter approach could undoubtedly be important to the study of hydration but in order to limit the scope of this review, we will not consider hydration in binary solvent systems and thus we are limited to gas phase isotherm studies.

The first step in the characterisation of micron-sized particles is frequently a surface area measurement obtained from the adsorption isotherm of gas phase nitrogen at 77 K using the isotherm originally described by Brunauer, Emmett and Teller (BET)[4]. The model underlying this isotherm is that of multi-layer adsorption with each layer in equilibrium with the vapour but with only the first layer having an energy of adsorption different from the liquefaction of pure solvent. The original theoretical development used kinetic arguments but more rigorous statistical mechanical developments also now exist[5,6]. It should be mentioned that the BET isotherm is able to model the different behaviour in adsorption isotherms reasonably but it is in no sense unique as a theory. However, we are fortunate that virtually all researchers in the field use the BET area for nitrogen at 77 K as the standard measure for surface area which allows ready comparison of different solids studied by different groups even though the quantitative interpretation of the numbers may not be possible. Determinations are frequently made from a simple, two point adsorption measurement and commercial instrumentation is readily available.

Clearly, even within the terms of the BET isotherm, the value obtained for a surface area depends on the value chosen for the surface area of the adsorbing molecule. In certain instances, on very well-defined single crystal surfaces, the packing of molecules can be observed directly by low energy electron diffraction (LEED)[7]. Thus, at sufficiently low temperatures nitrogen has been observed to pack in a characteristic herringbone pattern on graphite, presumably due to the dumb-bell shape of the molecules. However, the herringbone separation appears to be mediated by the lattice spacing on the surface of the solid. At higher temperatures, different, less ordered packing is seen until eventually, nearly all orientational order is lost. There is some computer simulation evidence (discussed below) that this is the case. At higher temperatures, all of the orientational order is lost. As we change from one adsorbing molecule to another, the problem of defining the area occupied by one molecule becomes even more severe. Water adsorption often leads to isotherms that do not follow the BET theory and surface areas are then evaluated (albeit approximately) from the start of the linear portion (if it can be identified) beyond the characteristic knee or 'B' point in the isotherm.

In more general terms Brunauer[8] characterised different adsorption types in terms of five characteristic shapes and these are shown in figure 1; it is general practice to discuss complete isotherms in terms of these different shapes. When the heats of adsorption are greater than that for liquefaction of pure water we typically see type II behaviour, but for heats less than that for liquefaction, type III behaviour is observed. Surface areas for water adsorption less than those for nitrogen are frequently discussed in terms of surface hydrophobicity. Type III isotherms would also suggest strongly hydrophobic surfaces. Hydrophilicity is also sometimes discussed in terms of

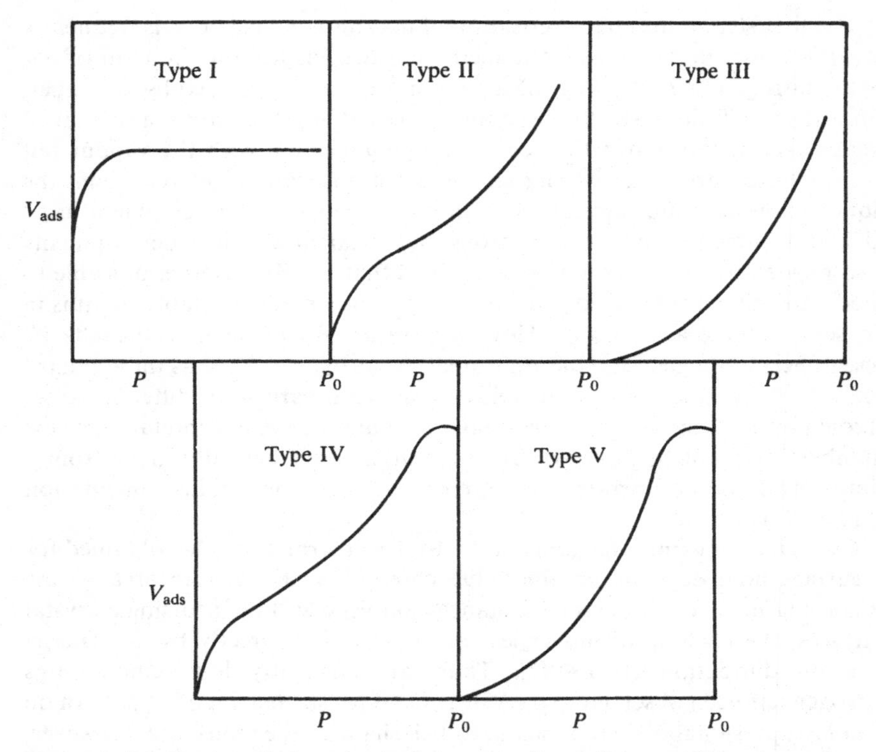

Figure 1. Classification of adsorption isotherms according to Brunauer. The volume of gas adsorbed V_{ads} is plotted as a function of applied pressure P up to the saturated vapour pressure P_0. Type I, Langmuir isotherm limited to monolayer coverage. Type II, Strong physical adsorption followed by multi-layer adsorption. Type III, Weak interaction between surface and water followed by multi-layer, adsorption. Types IV and V, Capillary condensation which results in effective reduction of area available for further adsorption. (Redrawn from ref. 1.)

the fractional coverage at the 'B' point. Theoretical treatments of isotherms which concentrate more on the multi-layer region have also been developed and the interested reader is referred to these studies[9].

The temperature dependence of adsorption isotherms relates through the Clausius–Clapyeron equation to heats of adsorption and it is convenient to include such effects in the following section on heats of immersion in solids.

A technique related to that of isotherm measurements which is crucial to our understanding of hydration phenomena, particularly on oxide surfaces which are highly heterogeneous, is that of temperature-programmed desorption. The technique is very straightforward, involving simply the heating of a prehydrated solid material at a carefully controlled rate and either analysing the mass loss of the material or perhaps analysing the quantity and chemical nature of the evolved gases as a function of temperature. Weakly physisorbed water is generally desorbed at relatively low temperatures whereas certain types of chemisorbed water are lost at only very high temperatures. For example, adjacent MOH groups can be readily condensed to yield a M—O—M bond and free water, whereas isolated OH groups are in general only lost at a temperature high enough for them to become mobile on the surface so that they may condense together.

2.2. *Heats of immersion*

Heats of immersion prove to be a most powerful technique for studying the interaction of water with solid material[10] and (at least on first glance), surprisingly, can be used to quantify the interaction of solids with *gaseous* water molecules. This arises because one may compare the heat of immersion of the clean, dry, outgassed solid with that of the solid with a known amount of water preadsorbed and thus by difference obtain the heat for the gas phase adsorption process (it is necessary to include in the calculation the heat of liquefaction for the number of moles of water preadsorbed on the solid). The quantities generally available from such calorimetric experiments are the total heat of immersion and the integral heat of immersion up to the vapour pressure used for the gas phase preadsorption. The temperature dependence of the adsorption isotherm by contrast yields what is referred to as the isosteric heat of adsorption. Suitable integration of this latter quantity with respect to coverage (in mol cm^{-2}) yields a heat which relates to the integral heat, as obtained calorimetrically. The detailed algebra connecting the various heats is discussed by Zettlemoyer, Micale and Klier[1]. Typically, heat of immersion isotherms follow one of five types[10], which are shown in figure 2. Homogeneous surfaces exhibit a high initial value for the heat of immersion that decreases linearly with coverage, eventually reaching a plateau at the value for a fully wetted surface. The linear portion arises because it is only at relatively high coverages that water–water interactions become important. Heterogeneous surfaces show an exponential decrease of heat with coverage (i.e., no linear initial decrease) due to immediate formation of water clusters around the active site. Most oxide surfaces exhibit this behaviour in water. Hydrophobic surfaces show a low initial heat due to the small interaction between the surface and the water and it is only at relatively high pressures that water clustering occurs with concomitant rise of the heat. Clay materials and other layer minerals capable of intercalating water molecules (see below), exhibit a stepped behaviour since intercalation only occurs above certain critical pressures. Finally, porous solids exhibit lower values for heats of immersion at high pressures than expected, due to filling of the pores which reduces the effective surface area.

Study of the variation of heats of immersion with the temperature of pretreatment (activation) of the surface is a particularly useful approach

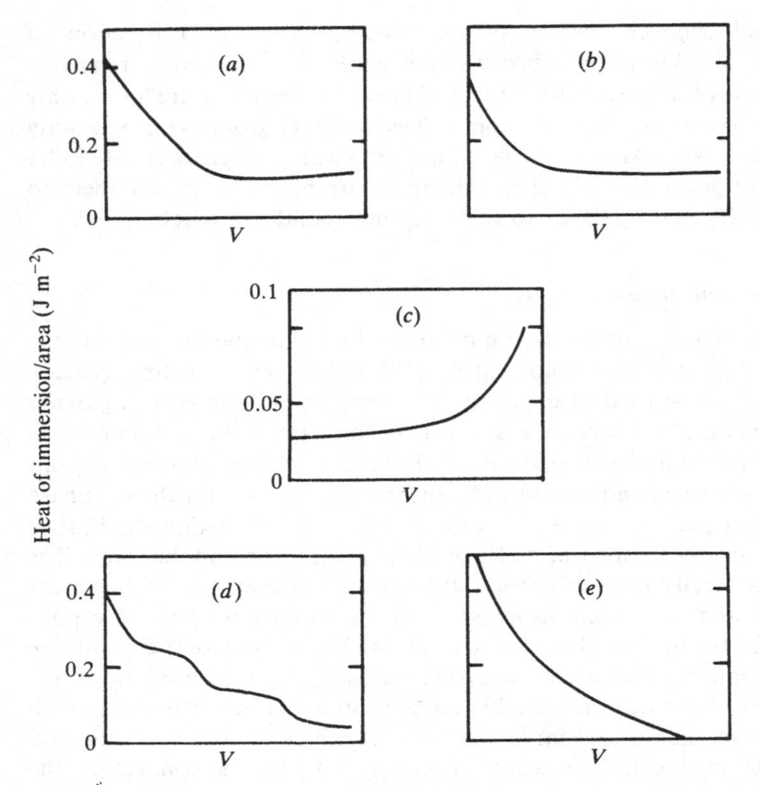

Figure 2. Classification of heat of adsorption isotherms: (*a*) homogeneous surface; (*b*) heterogeneous surface; (*c*) hydrophobic surface with a few hydrophilic sites; (*d*) clay or other layered material; (*e*) capillary condensation. (Redrawn from ref. 1.)

towards understanding the process of hydration of oxide surfaces[10]. For low temperatures of surface activation the enthalpies are relatively low, reflecting loss of only physisorbed water. For higher temperatures, chemisorbed water is lost with subsequently higher heats of rehydration. At very high activation temperatures, the heats may decrease if, as in the case of silica (see below), the reactive hydroxyls are converted to relatively inert siloxane bonds. Similarly, the Clausius–Clapyron plots for water adsorption on titanium dioxide in the temperature range 25–200 °C show a discontinuity at low coverages which has been attributed to reversible chemisorption at high temperatures[11].

The wealth of information obtainable from both isotherm measurements and heats of immersion is clearly considerable and they have been instrumental in developing our understanding of the interaction of water with surfaces. However, as is frequently the case with thermodynamic measurements, the detailed interpretation of such results in terms of the intermolecular interactions is extremely difficult and site-specific probes (generally spectroscopic) rather than macroscopic bulk measurements are crucial in order to elucidate the details of the interactions.

2.3. *Computer simulations*

Computer simulation methods are included within the section on experimental methods since, in principle, if the correct intermolecular potential energy functions are used and the size of the simulation is large enough in terms of numbers of molecules and length of time of simulation, then the simulation results should accurately model the system. Thus one is able to obtain pseudo-experimental results of a structural, thermodynamic or dynamic nature which may be difficult if not impossible to obtain directly from experiment. Computer simulation methods have had a profound impact on our understanding of the structure and dynamics of liquids and it is inevitable that similar methods should be used to study inhomogeneous systems, in particular interfacial regions. Meaningful simulations of such regions are extremely expensive in computing time; the necessarily small fraction of molecules within the interface region makes it difficult to obtain adequate statistics; it is also necessary to choose an ensemble large enough to model both of the pure phases; solvent molecules move very rapidly and yet any mass transfer effects occur on a very much longer time scale, implying (for molecular dynamics (MD) calculations) very short integration steps and yet very long equilibration times or (for Monte Carlo (MC) calculations) very large sampling sets. However, if future supercomputer advances continue at anything like the rate of advances in the recent past, it can be safely predicted that simulation approaches will become increasingly popular. However, a major problem yet to be addressed in this area relates to the very nature of heterogeneous systems. Homogeneous systems can be modelled due to their very homogeneity; in contrast, experimental results for heterogeneous systems are often dominated by surface impurities, imperfections or even morphological effects such as porosity and surface roughness. As a simple but yet telling example of this, we may note that magnesium metal can be burnt in air or oxygen to produce a white smoke of magnesium oxide which consists of highly perfect micron-sized single crystals[12]. These crystals (according to infrared absorption data) adsorb water vapour at a negligibly low rate. However, if the same crystals are ground very, very lightly between two glass plates, the resulting solid adsorbs water vapour from the air extremely rapidly. Electron microscopy suggests that the only damage to the crystals involves removal of their sharp corners. Clearly, modelling the possible defect and crystal imperfection sites which may dominate certain surface properties is no trivial matter.

MC methods[13] involve setting up the atoms and molecules in the hypothetical system at certain random coordinates and calculating the energy of the ensemble from the various intermolecular potentials. One molecule is then moved a short distance away from its original position at random and a new value for the energy is calculated. This new configuration is rejected if the new energy is higher than the old or accepted according to a Boltzmann weighting if it is lower. The process is iterated an extremely large number of times so as to yield both structural and thermodynamic properties. MD simulations[13] involve the numerical integration of Newton's equations of motion for a set of particles (atoms, molecules etc.) which are subject to certain interatomic forces and are given initial random velocities. Typical integration time steps are 10^{-15} s or less. The system is allowed to 'equilibrate' for a certain time (typically 10^{-11} s) after which the simulation is continued but during which time both structural, thermodynamic and dynamic parameters can be calculated. The MD and MC calculations which are inherently different in approach will be treated together in this section since both yield similar sorts of information; the advantage of MD calculations is that they yield dynamic as well as structural information which under favourable circumstances can be compared with direct experimental results. Condensed phase MD and MC studies of heterogeneous systems have included the surface of molecular liquids[14], ionic crystal surfaces[15], as well as gas–solid and liquid–solid interfaces. The gas phase simulations have only recently been reviewed and will therefore not be treated here[16]. The ionic crystal surface results which are relevant to this review suggest that for the 100 and 110 surfaces of alkali metal chloride crystals, there is a net increase of up to 4% in interlayer separation close to the surface with normal values of the separations occurring only ten layers down into the bulk. Surface buckling was also observed for lithium and sodium chlorides, giving rise to large negative surface potentials. This effect presumably arises due to the large size differences between cation and anion in these salts. Interestingly the 110 surface of sodium chloride was observed to be unstable at temperatures close to the melting point.

MD modelling has also been undertaken on *vitreous* silica surfaces[17]. The simulations assume that silica can be represented as an array of Si^{4+} and O^{2-} ions and the system is allowed to equilibrate at 10000 K (above the melting point of silica). An interface is then created and the temperature is rapidly reduced to 300 K and the system is allowed to reequilibrate (vitrify). Considerable surface rearrangement takes place during the computer quench so as to produce an excess of oxide close to the surface, non-bridging oxygens and strained siloxane (Si—O—Si) bonds. Three-coordinate silicon which is initially generated during the formation of the interface is generally removed during the surface construction. Edge sharing SiO_4 groups are also produced as well as five-coordinate silicon. Assuming that the non-bridging oxygens, the strained edge defects and perhaps the five-coordinate silicons would be available for hydroxylation in a water atmosphere, the estimated surface density of hydroyl groups agrees well with experiment. Similar simulations have been made for lithium and sodium silicate glasses[18]. The results suggest that there is a surface excess of cations for the sodium but not for the lithium glasses in agreement with ion-scattering spectroscopy; this also agrees with surface tension expectations in $SiO_2 + M_2O$ melts.

Interfacial water can be modelled simply as a very thin 'free standing' film or the simulation can include the interface explicitly. Thus for example, a 1.3 nm thick water film with a density of 1.5 g cm^{-3} has been modelled by MC[13]. The density as a function of distance from the surface does not show the large oscillations and maxima close to the surface frequently seen in other studies (see below) where the solid is treated explicitly; the pair radial distribution functions g_{OO}, g_{OH} and g_{HH} all show considerably more structure close to the surface than in the interior of the film. In addition there is some dipole ordering close to the surface with the oxygen atoms oriented preferentially towards the exterior. Studies where the solid surface is treated explicitly can vary widely in complexity, ranging from the treatment of the interface simply as an impenetrable structureless wall all the way through to an array of atoms or ions of structure chosen to represent real solids. For these latter cases computer size and time constraints often pose severe limitations on the statistical accuracy obtainable. Models of surfaces as structureless walls which can be variously infinitely repulsive, soft repulsive or include attractive and repulsive terms in a Lennard–Jones type potential have been treated by several groups[19–26]. Some have concluded that there are large oscillations in density as a function of distance from the wall which only die away slowly in the bulk. Such effects have also been observed for other liquids close to walls; however, the tentatively emerging consensus is that such effects are due to incomplete equilibration of the system and are thus artefacts of the simulations. This is borne out by the fact that simulations performed on a thicker layer of water show very much reduced oscillations. The effect also appears to be sharply increased by increasing the liquid density (i.e., the applied pressure) and may be due in part to packing constraints of the solvent between the walls which for most simulations are less than 2 nm apart. Generally, the simulations report enhanced water structure closer to the wall, i.e., the first maximum in g_{OO} is increased. Most simulations suggest that the water molecules are oriented such that the oxygen is found preferentially closer to the wall. There is also some suggestion that the plane of the water molecule is oriented parallel to the wall. Reports of increased, decreased or similar reorientational motion of molecules close to the wall (with respect to 'bulk' water) have been reported. A MC simulation of a structureless hard metallic wall[27] suggests that the structural enhancement of water is even larger than for a non-metallic wall. The field is still somewhat confused; in studies of bulk liquid, it is relatively easy to calculate from the simulation, reliable values for various thermodynamic, structural or dynamic properties that can be compared with experiment and thus obtain insight into the deficiencies of the different simulations. Parameters within the simulation can be varied to assess their influence on the final results. In heterogeneous systems, comparison with experimental data is extremely problematical and the simulations are so expensive in computing time that systematic change of the many variables is not currently feasible. The interested reader is referred to the original literature for more details of this rapidly growing subject.

Simulations where the atomic nature of the wall is treated explicitly are a lot less numerous for obvious reasons. Anastasiou, Fincham and Singer[28] have performed a MD simulation of the realistic case of water in contact with a rigid-ion crystal surface corresponding to sodium chloride. The simulation involves 432 ions and 108 water molecules with an interfacial area corresponding to 18 ions of each charge. The results suggest a considerable ordering at the crystal surface with, on average, oxygen atoms lying above the cations and the O—H bonds pointing towards anions. The density of water increases very significantly close to the surface. The dynamic results also imply considerable ordering close to the surface. A silicate surface–water interface has also been modelled as a two-dimensional array of oxygen atoms similar to those in a phyllosilicate sheet[29] (this structure is discussed further below in the section on clay structure). The authors observe little difference for g_{OO} between bulk and surface water (this may very well reflect the fact that oxygen–oxygen distances are similar in bulk water and in the phyllosilicate sheet). However, surface water exhibits fewer hydrogen bonds than bulk water, exhibits more dipole orientation and also a much slower dipole–dipole orientation autocorrelation function implying more restricted reorientational motion. Realistic metal surfaces have also been modelled by various authors. Spohr and Heinzinger[30], in a MD study of water on platinum have observed large density oscillations close to the wall (but see discussion above); the g_{OO} radial distribution looks similar to the platinum surface and the characteristic peak at 0.47 nm due to the tetrahedral nature of water has disappeared. Parsonage and Nicholson[31] in a MC study of water on copper noted that the density oscillations were absent with a 4.1 nm thickness of water but there was evidence for the same at 2.3 nm. There was evidence for alignment of the molecular plane of water parallel to the metal wall and also for some weaker dipole orientation with the oxygen closer to the surface.

It is clear from the above discussion that the computer simulation of hydration is still in its infancy. However, the results available to date hint tantalisingly at the sorts of detailed surface information that could become available in the future. Assuming that the question of density oscillations close to the wall is simply one of artefacts in the simulation, questions that need to be addressed include the importance of treating the atomic nature of the surface; thus the work on water in contact with sodium chloride suggests that the ions and their surface arrangement are crucial in determining the structure of the water close to the wall. The work on metals and on the phyllosilicate sheet implies the same sort of conclusion. Clearly work with structureless walls cannot resolve the question and more work on different 'atomic' surfaces with varying chemical properties is required. With regard to structureless walls, work is needed to investigate the effect of softness of the repulsion term on the structure of the water and also the importance of attractive terms. However, on a more fundamental level, before such simulation results can be accepted, the method must be capable of reproducing the known properties of the system. To date, even for homogeneous aqueous solutions, this is not the case. For example, there is still considerable debate as to the best intermolecular potential function to use for water–water interactions since none of those proposed is capable of correctly reproducing all the properties of pure water. The problems are clearly compounded in aqueous solutions where there are several intermolecular potential functions to consider. In addition, demands on computing power, both in terms of speed and sample size, are extreme. The simulations that can be performed are very small both in terms of physical size and in terms of time; this gives rise to serious boundary problems and also causes the derived parameters to be extremely noisy. Supercomputer advances continue unabated but a full resolution of these practical problems really does demand significant increases in computer power.

In spite of all of these hurdles yet to be overcome, the amount of information available from a simulation is extremely large and includes radial distribution functions, angular correlation functions, dipole correlations as well as a whole host of time dependent correlation functions. From such functions, thermodynamic, dynamic and (sometimes) spectroscopic results can be calculated and compared with real experiments. At a more conceptual level, the ability to 'see' the arrangements of atoms and molecules on the surface and their evolution with time will surely enhance our understanding significantly. Thus, in spite of, or perhaps because of, all the effort that will be required, the results of the next generation of reliable computer simulations will almost certainly be extremely worthwhile.

2.4. *Dielectric studies*

The study of the dielectric properties of both homogeneous and heterogeneous systems is a mature and established subject[32]. In the area of hydration of solids it has had a major impact on our understanding of the orientation of the water molecules adsorbed and on their reorientational motion. The three related experimental approaches in such studies are: (1) the relative permittivity (or in pre-SI parlance, the dielectric constant), which is simply the ratio of the capacitance of a capacitor with the plates separated by the substance of interest and by a vacuum. The relative permittivity is, in general, determined by the numbers of polar molecules in the sample, their dipole moments and their relative orientations with respect to each other. The adsorption of water (as well as other polar molecules) onto solids rapidly increases the permittivity which reflects the utility of the method. (2) Dielectric relaxation involves the measurement of the relative permittivity as a function of frequency of the applied electric field (effectively all the way from DC to as high as tens or hundreds of gigahertz). As long as the frequency of the applied field is less than the characteristic frequency of motion (generally rotational but perhaps diffusional, see below), the molecule will contribute to the measured permittivity. Above such a frequency, the molecular dipoles are no longer able to follow the reversals in the applied field and thus the molecule is no longer able to contribute. In this way, the characteristic frequencies of motion are accessible. A temperature dependence allows the evaluation of the energy of activation for the motion. (3) The dielectric loss is the energy loss due to the orientation of the dipoles by the electric field. The response of a dielectric to an alternating electric field can be treated as a complex mathematical quantity involving both in-phase (real) and out-of-phase (imaginary) quantities. The out-of-phase component is the dielectric loss. The mathematics of the time dependence of random fluctuations is extremely complicated; however, under certain simplifying assumptions, the motion can be represented by a single time, called a correlation time, which is the characteristic cut-off frequency for the spectral density for reorientational motion of the dipole moment vector of the water molecule. However, in certain instances, it is necessary to postulate a spread of correlation times in order to fit the relaxation data. As with many techniques originally developed for homogeneous systems, the extension to the heterogeneous case can lead to severe complications and many of these problems have yet to be fully resolved[33].

So far, the discussion has concentrated on reorientational motion of molecular dipoles. However, if the sample contains free charge carriers, i.e., ions, then these will also interact with the electric field not by reorientational but by diffusive motion (conductance). Indeed, in conducting samples, the connection between dielectric response and conductance is intimate. In heterogeneous systems, in particular in porous materials and the like where the water is contained in three-dimensional networks, the water path may or may not be continuous depending on the relative humidity and the nature of the material. Extremely low frequency dielectric dispersions can be observed in such systems (e.g., zeolites and ceramics) and these have been interpreted in terms of frustrated or 'incomplete ionic conduction' of ions in the water micropools of incomplete water network[34,35]. The process is thus clearly related to percolation theory. The low frequency dispersions appear to be non-linear, i.e., there is not a linear relationship between polarisation and applied potential. Further insight has been obtained by studying the time domain dielectric response in the range 100 μs–10^3 s[36]. The authors concluded that some of the problems of analysing the higher frequency dispersion processes and the need to invoke correlation time distributions, may be due to difficulties in correctly subtracting the very low frequency dispersion. The importance of this low frequency process is only now beginning to be realised and much work remains to be done in order to quantify the effect.

Reliable experimental methods for dielectric studies have been available for many years[1,32] but it is pertinent to review briefly the area due to a number of relatively recent advances which have given the technique considerably more power. The classical method simply involves the measurement of capacitance in a capacitance bridge. The frequency range of this approach is typically 10^{-2}–10^{5} Hz. Above this frequency, inductance effects become increasingly problematical and resonant circuit techniques are used. Relaxation studies by such techniques can be extremely time-consuming since data could be required over the frequency range 1 Hz–10 GHz (9 decades). The method proposed by Fellner–Feldegg[37] involves working not in the frequency range but in the time domain and using the standard Fourier transform between the two. The technique, sometimes called time domain reflectometry or dielectric spectroscopy, consists of applying the electric field as a very short pulse of high frequency and detecting the time response of the sample. The pulse contains frequency components not only at the carrier frequency but also at frequencies in the range $\pm 1/2t$ above and below the carrier (where t is the pulse length). The method is in many ways analogous to Fourier transform NMR.

A totally unrelated technique[38] used in recent years with very great success involves the cooling of a sample (typically to 77 K) between two conducting plates which are charged to a high electric field (perhaps 10 kV cm^{-1}) so as to 'freeze in' the various dipolar polarisations. The electric field is then removed and the conducting plates are connected to a highly sensitive current measuring device. The temperature is then slowly increased at an accurately known rate; as polarisations become 'thawed', they are detected as a current flow across the electric plates. A detailed analysis of the current/time results, together with a knowledge of the heating rate allows the evaluation of the correlation times and activation energies for the various reorientational processes occurring. Information about the relative numbers of molecules involved in the different processes is also obtained. This technique has been used to very great effect in the study, for example, of the hydration of solid sugars and polysaccharides[39].

2.5. *Vibrational spectroscopy*

Infrared vibrational spectroscopy is also a mature subject and as such was one of the first forms of spectroscopy to be applied to the study of water adsorbed onto surfaces[1]. It is interesting to note that it is the very effects responsible for making infrared spectroscopy a difficult technique in the liquid phase (i.e., the very large absorption coefficients and hence the need to work with extremely thin samples) that make the technique so useful for water molecules adsorbed onto surfaces. Thus the technique is extremely sensitive allowing study of surfaces with only small amounts of water adsorbed; in addition, the spectral changes observed are frequently very large thus allowing subtle effects to be observed. Perennial problems with infrared spectroscopy have included problems of sample preparation which frequently give rise to major difficulties in quantification of spectra. However, the relatively recent introduction of Fourier transform infrared spectroscopy (FTIR), as well as enhancing sensitivity has also allowed ready computer manipulation of spectra, e.g., baseline correction, addition and subtraction of spectra etc. and thus permits much easier quantification; this major new instrumentation advance is thus giving infrared spectroscopy a major new impetus in the study of hydration of surfaces.

It is perhaps useful to discuss briefly questions of sample handling since they are crucial to success in this area[1]. Finely dispersed solid material in the form of fine particles will normally scatter incident light extensively and hence cannot be used; mulls with hydrocarbon oils give good transmission but the oil perturbs the surface. Potassium bromide disks also give good transmission but not only is it hard to vary the amount of adsorbed water but ion exchange with potassium bromide is possible. In general, disks pressed at high pressure in the absence of potassium bromide yield good results; thus for example, Rochester and coworkers[40,41] have used 25 mm diameter disks containing from 50 to 200 mg of solid in studies on silica, tin and iron(III) oxides (amongst others). Highly effective, high vacuum bakeout cells which also allow controlled addition of both solutions and water vapour have been developed to enable detailed study of such disks. Some concern has been expressed that the very act of disk formation at high pressure (typically several hundreds of atmospheres) may perhaps modify the surface structure[1]. Attenuated reflectance spectroscopy is an alternative to the pressed disk approach and is discussed in ref. 1. Metal surfaces by their very nature do not allow the penetration of the electric field of the electromagnetic radiation at infrared energies and so conventional techniques cannot be used. However, a simple reflectance technique, particularly when used in a Fourier transform spectrometer allows molecules both small and large (proteins) to be studied with relative ease[42]; in general a single reflection is all that is needed. The exclusion of the electric field from the metal also gives rise to a 'surface selection rule' that only molecules with their dipole moments normal to the metal surface will absorb. It is not clear how useful this technique will be in the study of the interaction of metal surfaces with water but it has been used in work on the adsorption of amino acids as well as various rust inhibitors.

The detailed interpretation of the infrared spectra of bulk liquid water is complicated due to both the complex nature of water and uncertainties in the detailed assignment of the many bands. Temperature dependence studies suggest that the bands frequently contain several components but deconvolution is not always easy. The interested reader is referred to one of the two chapters in '*Water, A Comprehensive Treatise*' by Walrafren[43] and Luck[44] on the power of the technique and of the wealth of information available. In order to give some idea of the utility of the technique in relation to surface hydration studies, it is perhaps useful to discuss representative studies of one of the most widely studied surfaces, silicon dioxide. Oxide surfaces frequently contain surface OH bonds which can also be studied in parallel. In the far infrared region ($< 800\ cm^{-1}$), one may observe librational bands caused by the restricted rotation of water molecules and these bands clearly show great sensitivity to changes in hydrogen bonding. Thus, for example, the highest frequency librational band in liquid water is $710\ cm^{-1}$ whereas in ice it increases to $810–850\ cm^{-1}$. In addition, the librational motion can also be studied by inelastic neutron scattering[45] (see below). Within the fundamental region ($800–4000\ cm^{-1}$), occur two bands, one around $1600\ cm^{-1}$ due to H—O—H bond bending (ν_2) and one intense, broad band around $3300\ cm^{-1}$ due to both symmetric and asymmetric vibrations (ν_1 and ν_3). Both bands are sensitive to hydrogen bonding changes in particular the latter. Any detailed interpretation of this latter band is complicated by the fact that the band contains contributions from overtones and combinations. It is frequent practice to decouple the hydroxyl vibration by working not in pure water but in an approximately 5% H_2O solution in D_2O. The $3300\ cm^{-1}$ band is due principally to hydroxyl centred vibrations and thus vibrational bands for surface hydroxyls occur in a similar region. Thus a free silanol bond gives rise to a sharp peak at around $3755\ cm^{-1}$ whereas hydrogen bonded silanol groups give a broad peak around 3600 cm^{-1}[46]. The physisorbed water gives a broad peak around $3300–3400\ cm^{-1}$. It is removed by outgassing at 383 K. The hydrogen bonded SiOH groups are removed by outgassing at 773 K (presumably by condensation) whereas the free SiOH vibration still exists after treatment as high as 1123 K. It appears that rehydroxylation of the surface to give hydrogen bonded silanol occurs after retreatment with water.

Within the near infrared region ($8000–4000\ cm^{-1}$) a series of harmonic and combination bands occurs notably at $6900\ cm^{-1}$ and at $5300\ cm^{-1}$. This latter band corresponds to $\nu_1 + \nu_3$. A free silanol gives rise to an overtone band at $7330\ cm^{-1}$ which is replaced by a broad band in the region $6880–7150\ cm^{-1}$ upon physisorption of water[47]. The combination band ($\nu_1 + \nu_2$) of physisorbed water is almost identical with that of pure water which strongly implies that the hydrogen bonded state of this physisorbed water is very similar to that of bulk water, i.e., the oxygen must be bonded to *two* silanols. It is worth noting that it is only in the near infrared region that one can unambiguously distinguish adsorbed water from silanol vibrations.

Raman spectroscopic studies are far less numerous due to the considerably lower sensitivity of the technique. However, recently the $3000–4000\ cm^{-1}$

region has been studied and closely similar results have been obtained[48]. In the area of surface–substrate interactions, it has been found that adsorbed species sometimes exhibit an enormously enhanced signal; this is referred to as surface-enhanced Raman spectroscopy (SERS). The general theory of the approach involves the coupling of the vibrational motion to an electronic transition of the substrate in a manner akin to a related Raman technique (resonance Raman). SERS has been reported on many occasions in the literature for the water–silver metal interface but it was always difficult to explain such an observation in terms of currently accepted theories. In addition, the enhancement of the signal correlated with the presence of certain anions in solution. Recently, however, it has been suggested that extremely minute carbonaceous impurities on the surface of the metal (as evidenced by Auger spectroscopy) are the cause of an enhanced Raman signal[49], and if these are rigorously excluded so as to give no Auger carbon peak, then no enhanced Raman signal is observed. The carbon impurities are believed to come from the salts used in the different studies. Regardless of this conclusion, SERS may still be useful to characterise the metal–water interaction (even though it may be a contaminated surface) and work is continuing in this area on the effect of salts and on the assignment of the spectra[50–2].

2.6. *Optical spectroscopy*

Most other forms of optical spectroscopy are not of much interest in the field of hydration of solids except for certain highly specific cases. There are two areas where the approach can yield important results. Thus for the oxides of the alkaline earth metals, both UV-visible and photoluminescence spectroscopy can yield crucial information about the surface and its reactivity. As discussed above[12], magnesium oxide crystals, obtained from burning magnesium ribbon in air, are highly regular and highly inert and they exhibit neither optical absorption in the UV-visible region nor photoluminescence [53]. However, the magnesium oxide prepared by heating the hydroxide or the carbonate yields highly imperfect, irregular crystals and these exhibit both UV-visible absorption peaks and photoluminescence peaks which can be identified with low (three- or four-) coordinate magnesium ions. The highly perfect magnesium oxide smoke crystals also start to exhibit the same spectral features as the imperfect crystals after exposure to water and they correlate well with electron microscopic observations of edge and corner erosion. Similar results have been obtained for other alkaline earths[54]. Clearly any technique that allows quantification of such defects could be of immense utility since much of surface reactivity relates to the density of these sites. Photoluminescence spectra appear to be more diagnostic than the absorption spectra (obtained as reflectance spectra) but the photoluminescence spectra are readily quenched by molecular oxygen and by close proximity to paramagnetic sites. Similar photoluminescence spectra have been observed in thorium oxides and both bulk and surface low coordination metal ion defects have been identified[55]. Dehydration luminescence has recently been observed during the thermal dehydration in kaolinite[56]. The effect is extremely weak (and hence hard to quantify) and the mechanism has yet to be elucidated.

The other area where UV-visible spectroscopy has an impact is with solid materials such as clays or zeolites which are able to undergo ion-exchange and are also able to incorporate water within their structures. Thus, if a transition metal ion is ion-exchanged into the structure, successive dehydration frequently gives rise to a reduction in the hydration number of the ion; this is readily monitored by optical spectroscopy.

2.7. *Magnetic resonance*

Nuclear magnetic resonance (NMR) has had a profound impact on our understanding of the nature of both homogeneous and heterogeneous aqueous systems. The scope of studies for the surface–solvent interface continues to grow as high pulse power spectrometers become more readily available and 'solid-state' studies become as accessible as well as the more well-known 'liquid-state' studies. NMR is ideally suited to studies of reorientational and diffusional motion. The range of nuclei available is very large and embraces all water species (^1H,^2H and ^{17}O) as well as a plethora of possible nuclei in the solid itself. To date, virtually all the NMR studies dealing with the hydration of surfaces have dealt with water nuclei. Earlier work in this area is well reviewed by Derbyshire[57]. However, in principle, nuclei in the solid close to, or at, the surface should show vastly different NMR properties from those in the bulk particularly if in the near vicinity of a (relatively mobile) water molecule. The major limitation will obviously be the relatively few atoms at the surface compared to the total number in the solid. We confidently expect that studies of nuclei in the solid phase will become increasingly important in the future but in this review we will concentrate on water nuclei.

2.7.1. *Chemical shifts*

In conventional organic chemistry NMR, the use of chemical shifts to elucidate structure, reigns supreme whereas for surface studies its utility is much more limited. The NMR chemical shift is a measure of the magnetic field experienced by the nucleus and as such is affected by the electronic structure of the molecule due to magnetic shielding by the electron cloud. It is also pertinent to note that the chemical shift will change if different parts of the sample experience different magnetic fields. Thus, in conventional high resolution work, corrections have to be made for the case of the chemical shift reference being contained in a separate capillary if the bulk magnetic susceptibilities of sample and reference solution are different. In a heterogeneous sample, the situation is very unclear since, if the solid material has a different magnetic susceptibility from the liquid, then molecules at different distances from the solid would experience different magnetic fields; in addition, solvent molecules would be in inhomogeneous fields and would give rise to line broadening.

In principle, we should expect that each individual chemical or magnetic environment should give a distinct NMR resonance. However, if the exchange between the various sites is faster than the reciprocal of the chemical shift (in rad s^{-1}) between the various sites, then only a weighted average chemical shift will be observed. With the exception of protons close to a paramagnetic species (transition metal ion or perhaps a free radical) where large chemical shifts are possible, the chemical shifts of water molecules in different environments will rarely exceed 2 ppm. Thus, for example, with a typical modern superconducting solenoid spectrometer operating at 250 MHz, we cannot resolve different water signals if the proton exchange is faster than 0.3 ms. In general, proton exchange rates between different water molecules are many orders of magnitude faster than this. For ^2H the situation is worse since for any given spectrometer, the chemical shifts (in hertz) are only ~ 15% of those of ^1H. For ^{17}O the relevant exchange rate is that for the whole molecule and chemical shifts are also larger than for protons. However, linewidths are often very large, ^{17}O is not very abundant (0.03%), it is not a very sensitive nucleus and isotopically enriched water is extremely expensive. As mentioned above, in the presence of very fast exchange, the observed chemical shift δ is the weighted sum of the hypothetical chemical shifts of each environment. Thus we may write for an exchange between sites A and B:

$$\delta = x_A \delta_A + x_B \delta_B \tag{1}$$

where x_i and δ_i are the mole fractions of site i (i = A,B). For electrolyte solutions, the two sites could be 'free' water and 'bound' water. A major problem with such studies is that both the chemical shift and the mol fraction of 'bound' water are unknown. The 'bound' chemical shifts are affected by both the electric field of the ion and hydrogen bond changes which occur upon hydration and considerable success has been obtained in interpreting such shifts in spite of the problems noted above[58]. Similar arguments would apply for water connected with surfaces although the ability to perform such measurements depends on whether the chemical shifts are larger than the line broadening that invariably accompanies the partial immobilisation of a molecule on a surface (see below). Chemical shift changes have been observed for water in the pores of controlled pore glasses[59]. For a 7.5 nm pore glass, the chemical shift changed approximately 1.6 ppm upfield on going from pure water to approximately 4% water by weight in the glass. If it is assumed that susceptibility effects may be neglected (which is by no means clear), then the results imply that adsorbed water is less hydrogen bonded than pure water. Proton NMR shift measurements of ammonia on graphite show very large upfield chemical shifts (up to 30 ppm) which are thought to be caused by the very high anisotropy of the magnetic susceptibility of graphite[60]. The effect is akin to the ring current effect observed in benzene where either large downfield or upfield chemical shifts are observed depending on the position of the nucleus relative to the aromatic ring. In this case it can be concluded that ammonia is above the graphite basal plane.

2.7.2. *Nuclear magnetic relaxation*

NMR involves the use of a permanent magnetic field to split the degeneracies of the various nuclear spin energy levels and the application of a fluctuating magnetic field (generally at radio frequency energies) to stimulate transitions between the levels. After excitation, the probability of spontaneous return to the ground state is negligible and to effect such return, the nucleus must couple to a random magnetic field fluctuating at the resonance frequency. This stimulated return to the ground state is known as relaxation. There are different relaxation times in NMR but they all exhibit the same general form[61,62]

$$1/T_i = K_i\overline{H^2}f_i(\tau,\omega) \qquad (2)$$

where T_i is a relaxation time, K_i involves certain fundamental constants, $\overline{H^2}$ is the mean square fluctuating field and f_i is the relaxation function and is a function of both the resonance frequency ω and a correlation time τ which is a measure of the time scale of fluctuating field. For highly fluid systems, such as liquids and solutions of small molecules (less than $\sim$ 1000 molecular weight) where $\omega\tau \ll 1$, the relaxation function equals the correlation time and all the different relaxation times are approximately equal and separation of the terms τ and $\overline{H^2}$ becomes problematical. For systems exhibiting motion slower than the condition $\omega\tau \gg 1$, the various relaxation functions f_i become different and exhibit highly specific frequency dependences thus allowing separate evaluation of τ. Water molecules are frequently sufficiently immobilised so that the condition $\omega\tau \gg 1$ is fulfilled. A major limitation of nuclear magnetic relaxation relates to that discussed in the section on chemical shifts. If the exchange between two sites is sufficiently fast that only a weighted average peak is observed, the relaxation times measured are also the weighted averages of the two environments. As well as the problem of not generally knowing the fractions in each site (generally treated simply as bound and free), if the fraction of bound molecules is very small, i.e., we have a lot of bulk water with only a small fraction bound, then the effect we observe can become immeasurably small. Water vapour–surface interactions where there is no bulk water are, of course, immune to such problems.

One can identify three possible relaxation mechanisms likely to be important in the NMR of surfaces. (*a*) Dipolar relaxation[61] – in this relaxation mechanism which is almost certain to be predominant for ^{1}H of water, adjacent nuclear dipoles provide the fluctuating magnetic field. The large value of the gyromagnetic ratio of ^{1}H dictates that proton–proton interaction will be by far the most important. However, as in the case of bulk water, both inter- and intramolecular effects are likely to be important; for solvent molecules without proton exchange, dilution with deuterium substituted solvent can allow resolution of this problem but this is not so for water where proton exchange is rapid Unpaired electrons either on free radicals or on paramagnetic transition metal ions have gyromagnetic ratios very much larger than the proton (for the free electron, it is 658 times); this parameter comes in as a squared term in the relaxation rate and the presence of paramagnetic species on a surface has a profound effect on the relaxation rates observed. Frequently, the presence of such impurities in minute quantities can destroy the utility of the NMR relaxation technique. (*b*) Quadrupolar relaxation[61] – this mechanism is possible only for nuclei with a spin greater than $\frac{1}{2}$ (e.g. ^{2}H $\frac{3}{2}$; ^{17}O $\frac{5}{2}$). Under a classical picture, a spin greater than $\frac{1}{2}$ implies a non-spherical distribution of charge on the nucleus; in the presence of an electric field gradient across the nucleus, they precess about the nett electric field and in doing so provide a relaxation mechanism[62]. (*c*) Chemical shift anisotropy[61] – the chemical shift is strictly a tensor quantity although in solution the rapid tumbling causes the measured value to be simply the trace of the tensor. However, the tensor nature of the chemical shift implies that the shift value (a measure of the magnetic field at the nucleus) depends on the orientation of the molecule with respect to the applied magnetic field. Molecular motion thus causes a fluctuating magnetic field which gives rise to relaxation. The interesting aspect of this relaxation mechanism is that the extent of the fluctuating field depends on the value of the applied field unlike the other relaxation mechanisms and, in general, it is easy to separate this relaxation mechanism from the others by comparing results at different applied fields. This relaxation mechanism is not expected to be very important for most surface studies. Spin rotation interaction[61] is in principle a fourth possible relaxation mechanism but it is generally only important for extremely non-viscous and hence rapidly reorientating molecules where dipolar relaxation becomes inefficient. It is thus important for gases and for fluids at high temperatures and low densities. One would not expect it to be important for adsorbed species except perhaps at extremely high temperatures and it will not be considered further here.

There are three different relaxation rates (relaxation rate is simply the reciprocal of the relaxation time) of interest in connection with hydration of surfaces. It is beyond the scope of this review to give a detailed discussion of the relevant NMR relaxation theory here and the interested reader is referred to one of the excellent texts in this area[61,62]. The most commonly

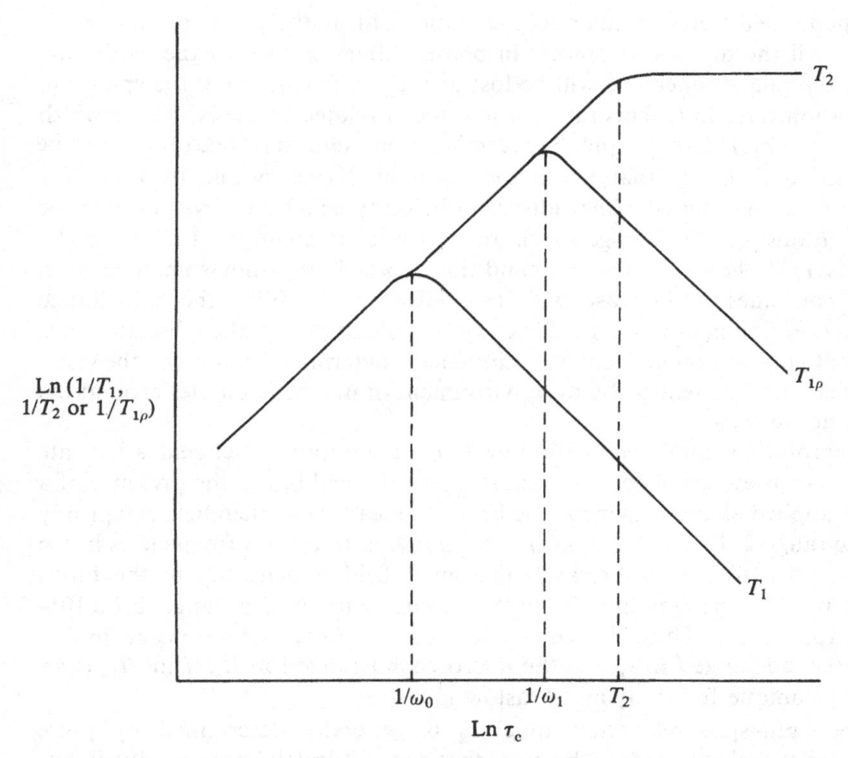

Figure 3. Schematic representation of variation of spin–lattice ($1/T_1$), spin–spin ($1/T_2$) and rotating frame spin–lattice ($1/T_{1\rho}$) relaxation rates with change of correlation time.

encountered rate, the spin lattice relaxation rate, $1/T_1$ is simply the first order rate constant for the return to the ground state of an excited state in the applied magnetic field. From the above discussion it is clear that fluctuations at the resonant frequency will stimulate the relaxation. A more rigorous quantum mechanical description also shows that for a system of two identical spins in close proximity (i.e., H_2O) fluctuations at exactly twice the resonant frequency are also effective. For a very viscous solution or considerably immobilised water molecules, the probability of fluctuations at a high enough frequency to cause relaxation will be small and the relaxation rate will be very slow. For a very non-viscous solution, there will be solution fluctuations up to very high frequencies; however, the intensities of these fluctuations are small and again relaxation is inefficient. Thus the spin–lattice relaxation rate shows a characteristic minimum as a function of correlation time at which point $\omega_0\tau \sim 1$ (figure 3). (ω_0 is simply $2\pi\nu_0$, with ν_0 the operating frequency.)

In the classical picture of NMR, the various nuclear dipoles precess in a magnetic field at the resonance frequency. Application of a small oscillating magnetic field perpendicular to the static field at the resonant frequency, causes all the dipoles to precess in phase. After removal of the oscillating field, the phase coherence will be lost at a characteristic rate, the spin–spin relaxation rate, $1/T_2$. This rate is also directly related to the NMR linewidth $\nu_{\frac{1}{2}}$ ($1/T_2 = \pi\nu_{\frac{1}{2}}$). Unlike spin–lattice relaxation, spin–spin relaxation can be induced both by fluctuations at the resonant frequency and by very slow motion fluctuations (the linewidth can obviously be affected by slow or static fluctuations i.e., inhomogeneity). In non-viscous solution, $1/T_1$ generally equals $1/T_2$; however, for the condition $\omega_0\tau \gg 1$, the spin–spin relaxation time continues to increase with correlation time unlike the spin–lattice relaxation (see figure 3). At sufficiently slow fluctuations, the relaxation rate ceases to depend on molecular motion and is determined simply by the static magnetic field generated by the environment of magnetic dipoles around the relaxing nucleus.

The rotating frame relaxation rate $1/T_{1\rho}$ is the spin–lattice relaxation rate not in the presence of the large static magnetic field but in the presence of a small applied alternating magnetic field. Whereas the applied field is typically in the range 2–12 T (20–120 kG) corresponding to proton frequencies in the range 5.4×10^8–3.2×10^9 rad s^{-1}, the small field is generally in the range 0.1–1.0 mT corresponding to proton frequencies in the range 2.7×10^4–2.7×10^5 rad s^{-1}. Thus the correlation times where $\omega_1\tau \sim 1$ occur in the nanosecond range for T_1 and the microsecond range for $T_{1\rho}$. Thus $T_{1\rho}$ is an ideal technique for studying ultraslow motions.

The spin–spin relaxation time T_2 is generally determined by pulse techniques unless one can be sure that the natural inhomogeneity is not contributing to the linewidth. If an artificial but known inhomogeneity, i.e., a field gradient, is added to the applied magnetic field, then it can be shown that the spin–spin relaxation rate determined by pulse techniques contains a contribution due to the field gradient and its extent is determined by the diffusion coefficient of the nucleus. Thus field gradient pulsed NMR approaches can be used to obtain diffusion coefficients of molecules[61,62].

2.7.3. *Static line broadening*

For an isolated water molecule rigidly fixed within a solid, each proton dipole exerts a magnetic field (which can be either positive or negative depending on its spin). On the other hand the resultant NMR spectrum consists of two lines whose separation (in magnetic field units) is given by $3\mu r^{-3}(3\cos^2\theta - 1)/2$ with μ the proton dipole moment, r the H—H distance and θ the angle between the H—H vector and the applied magnetic field. Clearly, an orientation study of water within a single crystal would allow detailed information on the orientation of the water molecule(s). In an amorphous solid where all orientations are equally possible, the resultant highly characteristic spectrum (frequently called a Pake doublet) allows evaluation of the H–H distance but all orientation information is lost. Similarly, if the protons are replaced by deuterons, dipolar effects are generally unimportant and quadrupolar effects dominate. Thus an isolated deuteron will give rise to a doublet due to interaction of its quadrupole moment with the electric field across the nucleus. The doublet separation (in frequency units) is given by $2*(3\cos^2\theta - 1)$ where $*$ equals $\frac{3}{8}$ of the quadrupole coupling constant and, assuming an axially symmetric electric field gradient, θ is now the angle between the ^{2}H–O bond and the applied magnetic field.

Rapid isotropic tumbling of the proton or deuteron at a frequency higher than the frequency separation of the doublet will average the peak to a single Lorentzian line. However, certain rotational motion is possible for the water that will not perturb the angular correlations. The approach can be extremely powerful particularly for solid materials where the quantity of water is not too large and thus where each water molecule is not too close to its neighbours. Thus, for example, both one[63] and two[64] water layer vermiculites have been studied by these techniques and the amount of both structural and dynamic information that can be obtained is remarkable. In the study on a two layer sodium vermiculite containing six water molecules per sodium ion, it is concluded that all six water molecules are associated with the sodium ion and are freely rotating around the Na—O axis. The $Na(H_2O)_6^+$ complex has considerable diffusional freedom, only being frozen out at low temperatures.

2.7.4. *Electron spin resonance* (*ESR*)

The prime requirement for the observation of an ESR signal is the presence of a species with one or more unpaired electrons and thus such studies, in general, would require the use of 'probe' molecules to assess the surface hydration. Such probes can, in principle, be organic free radicals[65,66] either charged or neutral, hydrophobic or hydrophilic, thus allowing different aspects of the surface to be examined. Alternatively, paramagnetic transition metal ions may be used most notably Mn^{2+} or Cu^{2+}, or the lanthanide Gd^{3+}. Thus for example Martini and coworkers have used both Mn^{2+}[67] and Cu^{2+}[68] to study the microdynamic nature of water in porous silica gels. It was observed that in narrow pore silica gels (4 and 6 nm pore diameters), the ESR spectra never exhibited frozen solution type spectra even down to 77 K implying that the pore water was non-freezable. By contrast, in wide pore silica gel (20–100 nm), at temperatures not far below the normal freezing point of water, the spectra appeared as the sum of frozen and non-frozen (glass-like) components. Correlation times for reorientational motion were estimated from linewidths. Detailed interpretation of such results is, however, severely hampered by the fact that one is, in general, observing the water of the hydration shell of the ion and this is not the same as the normal water in the system. One has no way of knowing what is the interaction of the ion with the surface and how it perturbs surface water. Also, the freezing of such solutions will lead to separation of salt and water and, assuming that all the salt stays in the non-freezing, bound water, Brüggeller has estimated the thickness of non-freezing water using ESR of Cu^{2+} ions in controlled pore glass[69] (see below).

2.8. *X-Ray, neutron and Electron scattering*

X-ray and neutron scattering methods are being increasingly used in surface studies and with the recent availability of synchrotron radiation sources to provide very intense monochromatic X-ray sources and also thermal neutron sources increasingly from ion (proton) accelerators rather than from nuclear reactors, the techniques should become increasingly important. Two reviews on the application of X-ray and neutron methods as applied to homogeneous electrolyte solutions should provide an introduction to some of the background theory and should also hint at some of the information that may become accessible in hydration of surface studies in the next few years[70,71]. The range of experimental approaches is quite large although they have not been used to the same extent as in homogeneous systems simply because of the added complexity of heterogeneous systems.

The classic X-ray and neutron diffraction methods have been applied to the two-dimensional surface structure of adsorbed molecules simply by subtracting the diffraction patterns of the clean solid from that of the loaded solid. Thus, for example, it has been shown that benzene adsorbs on the basal planes of graphite at low temperatures and loadings to give a two-dimensional monolayer with the molecules laying flat[72]. Above a monolayer coverage, benzene nucleates to form crystallites which have a structure similar to that of normal solid benzene. At higher temperatures (200 K), orientational freedom is observed. More recently, the structure of water within high pore silica gels has been studied by neutron diffraction[73]; this extensive paper discusses the additional problems encountered in the study of inhomogeneous systems and is a good introduction into the field. The scattering patterns of wet and dry silica gels were subtracted in order to give the scattering of the pore water itself. Such a procedure neglects atomic correlations occurring at the water–silica interface which may very well be important. The radial distribution functions (obtained by appropriate Fourier transformation of the scattering patterns) show broader peaks than for bulk water due to the finite size of the water sample. (The effect is analogous to the peak broadening observed in X-ray diffraction when the sample consists of very small crystallites.) However, differences in distribution functions for different temperatures for both bulk and pore water appear very similar, suggesting that the water structure is similar in both cases. The authors estimate that any effects due to differently structured interfacial water (see below; pore water) must be short-range and occur within 1.0 nm of the wall. For the narrowest pore silica (2.0 nm), there is just a hint of some enhanced O—H correlations. In a series of recent articles, Livage and coworkers[74–7] have used the combined techniques of electron diffraction, X-ray diffraction and neutron diffraction to study the vanadium pentoxide gel. Thus, electron diffraction was able to characterise the gel as being composed of randomly entwined ribbons approximately 100 nm × 10 nm × 1 nm in dimension. X-ray diffraction was able to identify the building blocks of the ribbon and the two-dimensional cell parameters. Neutron diffraction was able to quantify the enormous one-dimensional swelling that occurred on water uptake which appears to be stepwise for the first three layers and continuous beyond that. For example, the ribbon thickness increases to nearly 5.0 nm and the empirical formula reaches $V_2O_5 \cdot 20H_2O$ compared to only 1.6 H_2O for the gel dehydrated *in vacuo* at room temperature. Finally, low temperature neutron diffraction has revealed that apart from a small amount of water squeezed from the gel as the temperature is lowered and which shows a characteristic hexagonal ice structure close to 0 °C, the majority of the water (for a sample with composition $V_2O_5 \cdot 5–6H_2O$) was unfreezable. The enormous gain in information when parallel techniques can be used to study the same system under the same experimental conditions is self evident.

The advent of high power X-ray sources (in general electron synchrotrons) has meant that time-resolved studies of structures are becoming feasible. Thus, for example, studies have recently been made of the phase transitions in lipid bilayers following T-jump excitation across the phase change temperature. Thus 255 consecutive X-ray scattering patterns were obtained, each separated in time by only 150 ms[78]. The possibilities of this sort of approach not only in many different areas of surface chemistry but in other branches of both physics and chemistry are extremely exciting.

Conventional neutron and X-ray diffraction is sometimes referred to as wide angle scattering and for the conventionally used X-ray or neutron energies, the geometrical information that is collected is in the length scale of nanometres. If, however, diffraction (or scattering) patterns are collected at extremely small angles, the length scale can be enlarged to up to 100 nm. These modified techniques, generally referred to as small angle X-ray (or neutron) scattering, SAXS (or SANS), are thus able to give information about the size and shape of colloidal particles, micelles, vesicles etc. The observed scattering of a particle in water depends on the difference in scattering intensity of the particle and of the solvent. Each of these terms depends on the coherent scattering lengths of the constituent atoms; for neutron scattering, the scattering lengths of ^1H and ^2H are of different signs and it is possible to find a $^1H_2O/^2H_2O$ mixture where the total scattering is zero and thence to find the particle volume. This approach is known as contrast variation. The apparent radius of gyration of the particle is obtained from the angular variation of the scattering intensity and the correct radius value is obtained by extrapolation to infinite contrast. In general, the question of particle size is not of direct relevance to this review except for the fact that clay platelets (see below), can exist fully dispersed or in stacks of various sizes with interleaving water depending simply on the cation incorporated. Thus, for example, dilute sols of lithium montmorillonites exist fully dispersed, whereas the potassium sols exist as a pair of platelets and caesium sols as three platelets with two layers of water between[79]. In addition the contrast experiments showed that the proton and hence also water exchange between bulk and interleaved water was very slow.

In a solid material, neutrons may be elastically scattered from the atoms of the material or they may be scattered together with the exchange of a quantum of vibrational energy giving rise to a Doppler shift in neutron energy. In a liquid, Doppler shifts arise from exchange of energy with diffusional and rotational motions of the constituent atoms. The widths of the scattering peaks allow calculation of relevant diffusion coefficients. This technique is generally referred to as quasi-elastic neutron scattering (QENS). In heterogeneous systems, the situation is more complicated but is well understood. Thus, for example, in a two water layer clay, calcium montmorillonite[80] (see below), the diffusional motion of the water can be interpreted in terms of a Gaussian jump diffusion model with a root mean square displacement of 0.183 nm and a correlation time of 1.429×10^{-11} s at 300 K. However, the diffusional model is not unique to the data. Inelastic scattering spectra also yield similar information; a series of papers on the adsorption of ammonia on graphite shows the complementary nature of QENS and inelastic scattering, as well as the relation of these techniques to classical adsorption isotherms and their temperature dependence, neutron diffraction and NMR[81–4]. These four articles again show the power of using parallel techniques to understand the interactions of a surface with an adsorbing molecule.

2.9. *Other methods*

In the field of surface chemistry and physics, there are many new and exciting techniques for the characterisation of surfaces and of molecules adsorbed thereon. However, many of these methods rely on atomically flat surfaces and/or ultrahigh vacuum and are thus of perhaps limited use in hydration studies of real surfaces. However, in specific instances, such techniques can add information which is unavailable by any other method and thus for completeness, we should mention them here.

Electron microscopy is another standard technique for the characterisation of micron-sized particles and is crucial for the determination of their morphology. Scanning electron microscopy (SEM) typically has a maximum resolution of 10 nm. Frequently an X-ray fluorescence accessory is available which allows wavelength analysis of the X-rays produced by bombarding the sample with the high energy electrons, thus allowing chemical identification of the elements in various regions of the sample. Conventional transmission electron microscopy (TEM) has a higher theoretical resolution (0.2 nm). Both techniques require ultrahigh vacuum which clearly limits their use in hydration studies.

Tunnelling microscopy is a technique whereby atomically flat conducting surfaces may be probed so as to study the surface arrangement of atoms[85]. The theory behind the technique is very simple. When an atomically sharp needle is brought sufficiently close to the surface (typically less than 1.0 nm), and a small potential is applied between the two, a current will flow due to tunnelling of electrons across the gap; the current decays exponentially with the distance. Using piezoelectric positioners and paying particular attention to reducing thermal expansion effects to an absolute minimum, such devices are readily made and can be used to study the topography of atomically flat surfaces and also for the adsorption of macromolecules on the surface. The device does not require a high vacuum and has been made to work when the medium between the metal surface and the needle is water rather than air. It is perhaps too early yet to know how much impact this fascinating technique will have on our understanding of hydration of surfaces.

There are also a variety of other 'spectroscopic' techniques which are again ultrahigh vacuum techniques and are thus of only limited use in the area of discussion here. The incoming 'radiation' for these spectroscopies can be photons, electrons or ions as can be the outgoing beam[86–9]. The most common techniques are X-ray photoelectron spectroscopy (XPS) (often known as electron spectroscopy for chemical analysis, ESCA) where an X-ray (or vacuum ultraviolet) photon knocks out an inner shell electron whose energy is analysed to obtain chemical analysis of the elements on the surface of the sample. The energy resolution is high enough to allow one to distinguish different oxidation states for the element. In Auger electron spectroscopy (AES), an X-ray or energetic electron removes an inner core electron which is replaced by an outer core electron with simultaneous ejection of a further outer electron to restore the energy balance. Analysis of this Auger electron energy allows chemical identification of the atom. AES is of lower resolution than XPS but in the former technique, the exciting electron beam can be closely focused allowing lateral resolution of the sample and hence scanning. In addition depth profiling of the sample is possible by etching the surface with the electron beam. Both XPS and AES are more difficult with insulating materials due to charging of the surface; in addition AES, in particular, can be destructive. Of the 'ion' spectroscopies, secondary ion mass spectrometry (SIMS) and ion-scattering spectroscopy (ISS) are the most common techniques. In these two related techniques, a beam of inert gas ions impinges on a surface; this may either sputter secondary ions from the surface which are then analysed chemically by mass spectrometry (SIMS). Alternatively, the inert gas ions may be reflected elastically and their kinetic energy measured. Changes in energy depend on the mass of the atoms

Table 1. *Heat capacities of water near a variety of surfaces at 25 °C*[a]

Substance	Heat capacity (Cal g^{-1} K^{-1})
Bulk water	1.00±0.08
Porous glass	1.27±0.20
Activated carbon	1.28±0.03
Zeolite	1.21±0.03
Diamond	1.30±0.08
Collagen	1.24
Egg albumin	1.25±0.02
DNA	1.26±0.06
Artemia cyst	1.28±0.07

[a] See ref. 92 and references therein.

involved in the surface scattering. Ion bombardment techniques are surface destructive which allows depth profiling to be undertaken.

3. Pore water

It is useful to consider as a separate topic within the area of surface hydration, the topic variously called pore water, non-freezing water, vicinal water, non-specific surface–water interaction, in view of the intriguing results emanating from this area of research. This whole research area was considerably sidetracked during the 1970s by the 'polywater' episode which is now happily laid to rest[90].

In very simple terms, there appears to be considerable evidence that for a whole variety of surfaces which on the face of it would have vastly different interactions with water, the thermodynamic, spectroscopic and micro-dynamic properties of water close to the surface, depend simply on the distance from the surface. The names of Drost-Hansen and Derjaguin are synonymous with this area and earlier work in this area has been admirably reviewed[2,91]. As an example of the above idea, table 1 summarises estimates of the heat capacities of water at 25 °C close to a variety of surfaces. The results are very surprising in view of the very different surfaces studied[92]. As well as a higher heat capacity, surface water has a significantly lower density, higher expansivity, lower compressibility and a more positive value for the pressure differential of the apparent partial molar entropy[93]. Derjaguin, Karasev and Khromova have measured the densities of water in fine pores of silica gel[94] and titanium dioxide[95] and have shown that at high temperatures, the densities become identical and the anomalous effect is solely important at low temperatures (figure 4). The results in ^2H_2O, show a minimum in density for the pore water but shifted from the bulk value of 10.8 °C to 7 °C. It is thus reasonable to assume that the effect is related to the unusual properties of water at low temperatures. This is borne out by density

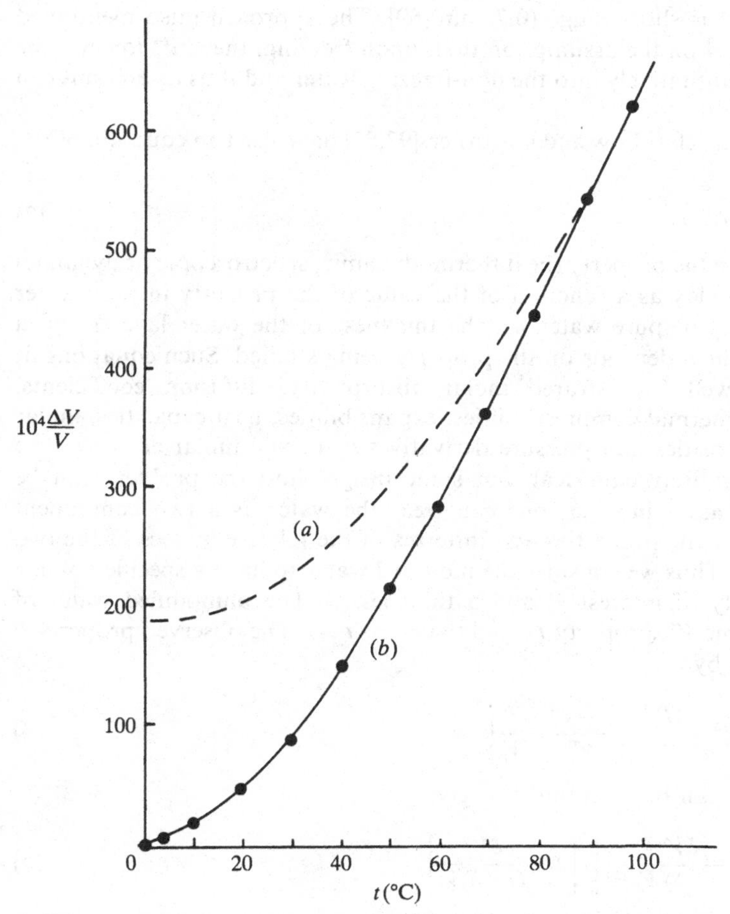

Figure 4. Differences in pore water volumes in (*a*) silica gel and (*b*) bulk water as a function of temperature. (Redrawn from ref. 94.)

measurements of various solvents in 14 nm pore diameter silica gels which are respectively for acetone, methanol and water in both pore (and bulk) 0.780 (0.785), 0.781 (0.788) and 0.966 (0.997) g cm^{-3}[96]. It is clear that the density reduction is only important for water. However, there is very little agreement as to the range of distance over which the modification of water structure occurs with estimates varying from 1 to 100 nm. The neutron diffraction study of water in high surface area silicas discussed above[73] is fairly specific in its conclusion that structural perturbation of the water has to be short-range in nature (1 nm) but the approach is of fairly low precision; in addition it is not certain the various properties involved in the phenomenon should necessarily have the same distance dependence. ESR measurements of the freezing behaviour of Cu^{2+} solutions in controlled pore glass also suggest that the effect is short range (0.72 nm)[69]. The approach (also mentioned above) is based on the assumption that, upon freezing, the Cu^{2+} ions will be expressed quantitatively into the non-freezing water and thus its amount can be estimated.

In several articles, Low and coworkers[93,97] have used an equation of the form

$$J = J^{\circ} \exp(k/\rho_w t) \tag{3}$$

to characterise the property (be it thermodynamic, spectroscopic or dynamic) J of water in clay as a function of the value of the property in pure water J°, the density of pure water ρ_w, the thickness of the water layer t and a constant k which depends on the property being studied. Such equations fit the results well for infrared molar absorptivity, diffusion coefficients, apparent isothermal compressibilities, expansibilities, heat capacities, molar volumes, viscosities and pressure derivatives of partial molar entropy. The equation is entirely empirical, but some insight into the problem can be obtained by assuming that one can treat the water as a two component mixture where the properties add in terms of the relative masses of the two components. Thus we consider the modified water to have a specific volume V_{δ}, a property of interest P_{δ} and a thickness δ. The unmodified water of specific volume V°, property P° and thickness $t-\delta$. The observed property P is thus given by

$$P = \left[\frac{(t-\delta)}{V^{\circ}} P^{\circ} + \frac{\delta P_{\delta}}{V_{\delta}}\right] \Big/ \left[\frac{(t-\delta)}{V^{\circ}} + \frac{\delta}{V_{\delta}}\right]. \tag{4}$$

Equation (4) can be rearranged to give

$$P/P^{\circ} = \left[1 + \frac{\delta V^{\circ} P_{\delta}}{(t-\delta)\, V_{\delta} P^{\circ}}\right] \Big/ \left[1 + \frac{\delta V^{\circ}}{(t-\delta) V_{\delta}}\right]. \tag{5}$$

By taking logarithms of both sides of equation (5) and (assuming that $t \gg \delta$) then we may expand out the logarithmic terms to yield

$$\ln\left(\frac{P}{P^{\circ}}\right) \approx \frac{\delta V^{\circ}}{t V_{\delta}} \left(\frac{P_{\delta} - P^{\circ}}{P^{\circ}}\right) \tag{6}$$

It is thus clear that the constant k in Equation (3) depends not only on the thickness of the layer but also critically on the fractional change in the property between its value in the bulk and close to the wall. From the results summarised by Sun, Lin and Low, the assumption required to perform the exponential substitution appears valid with the possible exception of infrared absorptivity.

Etzler and Fagundus[96] have endeavoured to perform a more general analysis of highly porous silica gels. The situation is different from the above case due to the different symmetry in the two cases. Thus, the mass of material close to a wall in a cylindrical pore of radius h is proportional to $2\pi h\delta/v$ whereas the mass of bulk material is proportional to $\pi(h-\delta)^2 V^{\circ}$ and this effect must be taken into account during the averaging process.

It is clear that our understanding of the above effect is far from complete; one would hope that at some stage in the future, computer simulations will further our understanding; however, to date, even the reduction of density of water close to the wall has yet to be reliably demonstrated. It is to be hoped that further work will explore the generality of the above effect and that as wide a range of experimental techniques as possible will be brought to bear on this problem.

4. Metals

One might expect metal surfaces to be a particularly fruitful area for study since a range of metals with a vast array of reactivities, structures and properties is available; single crystals are readily available allowing different surface arrangements to be studied. In practice nothing could be further from the truth. Controversy has raged for many years over even the most elementary question for metal surfaces; are they hydrophilic or hydrophobic? Schrader[98] has discussed in great detail the various attempts to resolve this issue for various metal surfaces and the heroic experimental precautions that it is necessary to take in order to solve the problem. As expected, it was concluded that surface purity is absolutely crucial to the correct determination of critical angles. Hydrophobic impurities at very, very low levels give rise to large angles. For gold, monolayer chemisorbed oxygen or a built-up oxide layer is not a problem but for other metals, such as silver and copper, where such effects occur, we might expect hydrogen bonding interactions with water molecules which would reduce contact angles to zero. Oxygen and oxide-free surfaces of these metals are problematical but there is now strong evidence that these metals are also hydrophilic. An additional complication is that a polishing treatment can introduce hydrophilic impurities on the surface and, additionally, that heat treatments can cause surface segregation of hydrophilic impurities. Thus, calcium was detected by AES on the surface of ion-bombarded and annealed 99.999+ % pure gold in both the presence and absence of oxygen. There is also some concern as to whether the contact angles of amorphous metal surfaces are lower than those of polycrystalline material or whether different crystal faces would give different contact angles; such effects were observed on graphite, but graphite is exceptional in having an extremely anisotropic crystal structure.

Contact angle measurements can be related to Hamaker constants.† These constants relate to the force exerted by a material based on pairwise summation of the dispersion force of atomic sized segments of the other

† The nature of the Hamaker constant is discussed in detail by Evans and Miller, see this volume, p. 27.

surface. The values obtained very early by Overbeek from gold sol stability measurements suggested large contact angles implying that other interactions than dispersion might be important. However, Hamaker constants have been obtained more recently directly from dielectric data alone using theoretical treatments of van der Waals forces based on a dielectric continuum model. These values imply zero contact angles. Of course, with surface impurity effects being so very important, these results may be of limited value in the 'real world'.

There is also very great interest in the interaction of ultrapure metal surfaces with water vapour at very high vacuum. Such studies generally utilise the high vacuum techniques such as LEED, ESCA or AES discussed above. Typical of such studies and also a useful review of the literature is the recent article by Hara and Debnath[99]. Pt(111), Rh(111) and Ru(001) adsorb water as a molecule through the oxygen lone-pair. However, on Fe(100) and Ti(001), water dissociates to give adsorbed H· and ·O—H. Ni(111) at 80 K adsorbs water and in the presence of preadsorbed oxygen, the adsorbed water is hydrogen bonded to the oxygen. At 120 K, a hydrogen abstraction process leads to the formation of adsorbed OH whereas above 200 K, H_2O is reformed from adjacent OHs and desorbs leaving behind adsorbed O:.

5. Oxides

For several reasons, oxides represent the most widely studied group of surfaces with regard to their hydration. The reaction of oxides is intimately related to the acid–base chemistry of aqueous solutions and distinct surface chemical reactions of surfaces with water molecules frequently occur (in surface chemistry parlance we speak of chemisorption). Oxide surfaces are thus frequently covered with acidic and basic groups capable of further reaction and/or strong hydrogen bonding interactions. In addition the number of chemical elements that give insoluble oxides is extremely large and thus extensive comparative studies on oxides with fundamentally different properties are feasible. From a technological point of view the corrosion of metal surfaces in aqueous solution is clearly related to the formation of oxides and their chemical and mechanical stability.

The studies on oxide surfaces are thus extremely numerous and, as well as involving the oxides of a large number of different elements, also involve many different experimental techniques and perspectives. In general terms we shall find that oxide surfaces in the absence of 'active' groups such as hydroxyls can be surprisingly inert. The surface concentration of hydroxyl groups is intimately related to the preexisting heat treatment of the surface (two surface hydroxyls will condense to form a water molecule and an X—O—X bond at high temperatures). The surface reactivity also relates to the kinetics of reactivation to regenerate hydroxyls (in normal chemical terms we would speak of hydrolysis of the surface). Some surfaces regenerate extremely rapidly whereas others are inert for extended periods of time. Much of the confusion in the literature relates to differences in surface structure due to different (or sometimes undefined) heat treatments. We should mention in passing that silica surfaces are considered to be heterogeneous with water molecules condensing on the active sites and eventually forming clusters of water molecules rather than an evenly distributed film of water. In addition, as mentioned above, differences in surface morphology and the presence of, for example high concentrations of low coordination number sites, will drastically alter the surface hydration. Silicon dioxide is by far the most widely studied oxide and as such it occupies a central place in our understanding of oxide surfaces. Thus, we shall consider it at length and our discussion of the other oxides will be made in comparative terms with the results of silicon dioxide.

5.1. *Silicon dioxide*

Silicon dioxide comes in a large variety of forms, the two common crystalline forms being quartz and crystobalite. In crystobalite, the silicon atoms are placed as are carbon atoms in diamond with the oxygen atoms midway between. In quartz, there are helices, so that enantiomorphic crystals occur. Interconversion of the two forms requires bond-breaking and is thus characterised by an extremely high activation energy. Very dense forms, coesite and stishovite, first made under extreme conditions but subsequently identified in geological material, are more chemically resistant than the normal forms. When crystalline silicon dioxide is heated above its softening temperature and slowly cooled it yields silicon dioxide glass, fused silica. A form of silica historically called amorphous silica is, in fact, microcrystalline silica finely ground. Fumed silica is an extremely fine non-porous powder with an enormous surface area (typical ranges: particle size 0.007 μm, area 400 $m^2\ g^{-1}$ to particle size 0.014 μm, area 200 $m^2\ g^{-1}$) and is made by gas phase hydrolysis of volatile silicon halides. Silica gels are amorphous hydrated silica $SiO_2 \cdot xH_2O$ formed from silicic acid by polycondensation of orthosilicic acid. The silicic acid sols have a micelle-like structure and some of this is retained in the gel giving rise to Si—O—Si (siloxane) bonds. Since the elemental particles touch, one produces a network, a conglomerate full of pores and capillaries in the range 30–100 Å. The heterogeneous surface is due to the various types of O—H bonds and surface siloxanes which may be formed. CNDO MO quantum mechanical calculations bear out the proposed structures and yield some estimate of their relative energies[100,101]; it is also worth reiterating that an ionic model of silica (i.e. Si^{4+} plus O^{2-}) appears to be a reasonable representation which has allowed the computer simulation of silica melts and (following a computer simulation of low temperature quenching) subsequent modelling of an anhydrous vitreous silica surface (see above)[17].

Both Si—O bonds and Si—C bonds are very stable towards hydrolysis and thus surface modification of silica surfaces has also been extensively used to produce solids with novel properties. Following initial reaction with either halo- or alkoxysilanes, extensive organic chemistry may be undertaken allowing, at least in principle, the production of surfaces with tailor-made properties. This area has had a major impact in the development of stationary phases in chromatography, in chemical synthesis using solid supports and in other areas.

A major consideration of hydration of silica relates to the surface concentration of hydroxyl groups. Experimental techniques may be divided into spectroscopic (generally infrared), isotopic exchange (with deuterated or tritiated water[102]) and chemical (with reaction with e.g. Me_2Zn). There is some concern that different techniques give different results[103]. It has been argued that the chemical method can give lower results if certain hydroxyl groups are inaccessible to the relatively large reagents employed; however, some of the reported differences may, in fact, be due to different sample handling procedures.

As mentioned above, silica surfaces devoid of hydroxyl groups are remarkably inert and behave as hydrophobic surfaces; they retain this property in damp air over extended periods of time. Normal silica generally contains about 5 hydroxyls per 10 nm. Heating to above 373 K will remove all physically adsorbed water whereas heating to above 773 K removes all hydrogen bonded, i.e., adjacent hydroxyl groups, while treatment at 1173 K removes virtually all hydroxyl groups[104]. The subsequently obtained isotherm is indicative of a hydrophobic surface although some vapour phase rehydroxylation occurs during the course of the experiment; complete rehydroxylation is not obtained.

Neutron diffraction and scattering experiments to elucidate the structure of water contained within the pores of silica gels and sols have been discussed in general terms above. The information obtainable is extensive and varied although the limitations of the different techniques may not yet be fully resolved; it is beyond the scope of this review to discuss the individual papers in great detail and the interested reader is referred to the original articles[73,105–8]. NMR, ESR, dielectric relaxation and neutron inelastic scattering have all been applied to the study of the microdynamics of water contained within silica gel or adsorbed on silica surfaces[68,69,109–12].

5.2. *Others*

Silicon dioxide turns out to be a somewhat special case in that high temperature dehydroxylation gives a surface which does not very readily rehydroxylate and is surprisingly hydrophobic. Many other oxides, after high temperature dehydroxylation, are subsequently completely rehydroxylated very rapidly at room temperature. The adsorption isotherm of such a dehydroxylated surface would obviously be non-reversible. It is perhaps worth repeating that highly perfect magnesium oxide crystals adsorb water very slowly; water adsorption is only rapid in the presence of low coordination number metal ions at surface defects.

The major oxides studied in depth to date are aluminium oxide, titanium dioxide, iron oxides and thorium oxide. For example, Rochester and Topham[113] have studied extensively the adsorption of water on haematite (α-Fe_2O_3) by infrared spectroscopy. Heat treatment must be performed in the presence of O_2 to prevent formation of magnetite (Fe_3O_4). Within the range 3000–4000 cm^{-1}, 11 infrared bands are observed whose relative intensities depend on the method of preparation of the haematite. Sintering also changes the number of hydroxyls per unit weight and the relative proportions of the different infrared bands. They argue that these variations in infrared intensities are due to changes in morphology of the haematite particles and hence to different relative areas of different crystal faces. Physisorbed water could be removed under vacuum at room temperature whereas surface dehydroxylation could be accomplished by heating, although full rehydroxylation occurred by subsequent water vapour addition at room temperature. Goethite (α-FeOOH)[114] gives rise to both surface and bulk hydroxyl infrared peaks which are distinguishable by ^2H_2O exchange. Both unperturbed and hydrogen bonded OH/H_2O vibrations could be observed. Decomposition to haematite at 185 °C *in vacuo* or 275 °C in oxygen prevented studies of dehydration or dehyroxylation. More recent FTIR studies of α-, β- and γ-FeOOH have also been undertaken[115].

Titanium oxide (rutile) surface hydroxyl concentration measurements (by tritium exchange) are in accord with the number of exposed Ti^{4+} ions on the surface[116]. The infrared results[117] show seven different bands in the 3400–3800 cm^{-1} region whose identities have been tentatively established. Some of the adsorbed water is only weakly held whereas other water molecules, identified as liganded onto coordinatively unsaturated, exposed Ti^{4+} ions, are strongly bound. Other more classical studies of rutile are also available[118].

Vanadium pentoxide has already been discussed above in connection with X-ray and neutron experimental techniques[74–7]. Other oxides studied include beryllium[119], aluminium[120,121], scandium[122], chromium [123,124], manganese and manganese 'nodules'[125], nickel[126], zinc[128], molybdenum[129], zirconium[130], thorium[131], plutonium[132] as well as various mixed oxides.

6. Insoluble salts

There is clearly a large array of insoluble inorganic salts other than oxides which have been extensively studied with regard to their surface hydration. Obvious candidates would include certain halide salts, most notably fluorides (due to the insolubility of some of their salts) but also silver iodide due to its use in cloud seeding experiments to produce rain. Clearly the interaction of the solid with water vapour is the crucial step in successful rain-making[133]. Other salts would include sulphates most notably with regard to concrete, Portland cement etc. Sulphides and phosphates also represent two classes of salts many of which are highly insoluble.

The computer simulations of water in contact with an insoluble sodium chloride-like salt suggest that the interactions of water molecules with the surface are very large[15] although such simulations are still in their infancy and it is not certain yet just how reliable they are. Infrared and isotherm results on various metal ion fluorides[134–6] suggest that the calcium and magnesium salts are different from the other salts studied (lithium, sodium, barium and lead) in that the latter only weakly adsorb water whereas the former strongly chemisorb either hydroxyls or water molecules directly onto the cations at approximately monolayer coverage. These hydroxyls/water are not capable of lateral hydrogen bonding. Subsequent layers of strongly physisorbed water also occur. The weakly physisorbed water for the latter groups of salts has been rationalised as the two hydrogens from a water molecule bridging two fluoride ions across the diagonal of the crystal lattice. It would be interesting to have MD calculations of the surfaces of these fluoride salts in order to see if this behaviour could be interpreted in terms of the buckling behaviour of the different salts[15].

Much of the work on the hydration behaviour of sulphates, most notably calcium sulphate, has been concerned with the morphological changes that occur during the various dehydration steps from the dihydrate to the hemihydrate to the anhydrous solid[137]. In addition two different hemihydrates are suggested to exist.

7. Geological materials

Certain classes of geological materials have highly developed molecular and morphological structures which are hard (if not impossible) to mimic in the laboratory; such materials are also frequently of enormous technological importance and can often have fascinating and very specific interactions with water molecules. Most of the structures are silicate materials and this is perhaps a justification for the enormous concentration on the properties of silicon dioxide itself. The two major groups of minerals are (i) the layered minerals of which the most important are the clays but which also include such materials as asbestos, talc and vermiculite and (ii) the zeolites. In this review, we will discuss the structures of both of these major groups together with their interactions with water.

7.1. *Clays*

The literature on the hydration of clays prior to 1978 has been admirably reviewed by Forslind and Jacobsson[3] and in this review we concentrate on the more recent work. We should also mention that the literature on clays and clay-like minerals is extremely large: there is, in fact, a journal, *Clays and Clay Minerals* solely devoted to this one area.

The general features of the structures of the clay minerals and the rules governing their molecular formulae were laid down over 50 years ago by Pauling[138]. However, their detailed geological classification is a highly complex field whose extensive nomenclature can be very confusing to the non-geologist due to the different names frequently given to minerals of closely similar structure. It is not the purpose of this review to discuss in detail the different structures of the various clays beyond pointing out similarities and differences in structures as they affect their varying hydration behaviour. The interested reader is referred to one of the standard texts in this area[139,140]. Clays are layered structure, aluminosilicate mineral particles with very small grain sizes ranging from 0.005 mm down to colloidal dimensions. They have a unique ability to intercalate water and other molecular species; in addition, by suitable chemical manipulation, the interlayer spacings of some clays can be increased quite significantly so as to accommodate either considerable quantities of water or else fairly large molecules. Clays are classified as phyllosilicate minerals meaning that they consist of linked SiO_4 groups each with three oxygens shared with adjacent groups giving Si_4O_{10} sheets; the apical oxygens always occur on the same side of the sheet. The other layer (sometimes referred to as the Gibbsite layer) consists of Al^{3+} ions sandwiched between two sheets of close packed oxygens (or hydroxyls) giving an octahedral coordination of the Al^{3+} ions. The different clay minerals vary with respect to the relative numbers of the two types of sheets and in the possibility of replacing the aluminium or silicon by other elements.

Kaolinites: The Kaolinite structure is the simplest of the clays consisting of one layer of silicate and one layer of Gibbsite and is shown schematically in figure 5(*a*). The ideal molecular formula is $Al_4Si_4O_{10}(OH)_8$. The *c*-axis repeat distance is 7.2 Å. Replacement of the aluminium and silicon by other elements is fairly rare. The different kaolinite clays arise from the different ways that the layers are stacked in the *c*-direction. Halloysite is a kaolinite clay with an expanded *c*-axis repeat distance of 1.0 nm due to intercalated water. It can be irreversibly dehydrated to give a *c*-axis repeat of 0.72 nm. Halloysite has a tubular structure due to a structural mismatch between layers (cf. chrysotile asbestos below).

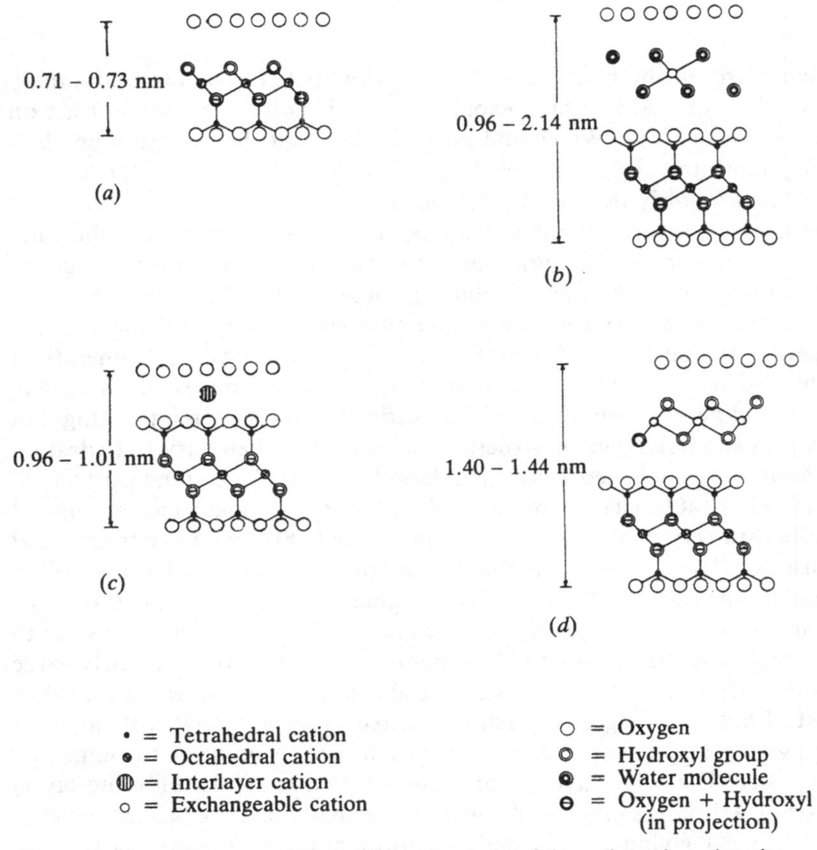

Figure 5. Schematic representation of layer structures of the various clay minerals: (*a*) kaolinites; (*b*) montmorillonites; (*c*) illites; (*d*) chlorites. (Redrawn from ref. 140.)

Montmorillonites: The montmorillonite or smectite group has one Gibbsite layer sandwiched between two Si_4O_{10} sheets as shown in figure 5(*b*). The ideal molecular formula is $Al_4Si_8(OH)_4O_{20}$. Successive sandwiches are stacked in the *c*-direction. The remarkable feature of this clay group is its highly variable *c*-axis repeat distance which varies from 0.96 nm in the fully dehydrated material to 2.14 nm in the fully hydrated state. Considerable substitution of both aluminium (by Fe^{3+}, Mg^{2+}, Zn^{2+}, Li^{+}, Cr^{3+}, Mn^{2+}, Ni^{2+}) and silicon (by Al^{3+}) frequently occurs. Such substitution will generally result in a nett negative charge (frequently about $\frac{2}{3}$ per unit cell). This charge deficiency is balanced by exchangeable cations (often Na^{+} or Ca^{2+}) adsorbed between the unit layers and around their edges.

Illites: The illite (or mica) group consists of clays which have the general structure of muscovite (see figure 5(*c*) which is similar to montmorillonite but with replacement of some (generally between 1 in 4 and 1 in 6) of the silicon by aluminium. The charge neutrality is preserved by inclusion of potassium ions between the layers. Muscovite does not exhibit the wide variability of *c*-axis repeat distances which is so characteristic of montmorillonite. The ideal muscovite structure is also the ideal structure of mica although well-crystallised micas generally exhibit more substitution of silicon by aluminium (generally 1 in 4) compared to the clays (1 in 6).

Chlorites: The chlorite clays (figure 5(*d*) are similar in structure to the montmorillonites except for the insertion of a layer (Mg, Al) (OH) between each montmorillonite layer and also by extensive replacement of aluminium by magnesium in the Gibbsite layer. There is, in fact, evidence for the formation of chlorite from montmorillonite in sea water which is rich in magnesium.

Allophanes: Allophanes are clays that are amorphous to X-rays (although they occasionally exhibit some weak lines). They consist of silicon in tetrahedral and metal ions in octahedral environments and occasionally PO_4; the composition varies widely and it is also hard to obtain pure samples. SiO_2/M_2O_3 ratios are frequently less than for the crystalline clays; allophanes can often only be postulated by negative evidence such as lower X-ray intensities than expected for a supposedly crystalline clay or else because the solvation or ion-exchange behaviour is inconsistent with the presumed crystalline clay.

We should perhaps add that the above clay structures are closely related to other important layer structure minerals such as talc, pyrophyllite, vermiculite and the two serpentine minerals, the fibrous chrysotile asbestos and the lamellar antigorite. Thus, pyrophyllite contains two adjacent Gibbsite layers sandwiched between two phyllosilicate sheets. Chrysotile has a structure related to kaolinite but with the Gibbsite layer replaced by a magnesium oxide (Brucite) layer except that all three sites are occupied by magnesium ions in order to preserve charge neutrality. The fibrous nature of chrysotile is caused by a certain structural mismatch between the layers which causes the sheet structure to be curved and thus a tubular 'swiss roll' structure is produced. Talc corresponds to pyrophillite with Gibbsite replaced by Brucite. Vermiculite is simply an alternation of mica with a double water layer.

The entire range of experimental techniques has been used to study the water within clays. The presence of cations not intimately connected with the lattice structure clearly impinges on the nature of the contained water since such cations would likely be hydrated. Thus, in the proton and deuteron

NMR static line broadening work discussed above[66,67], it is concluded that for the two layer sodium vermiculite containing six water molecules per sodium ion, all of the water molecules are associated with this cation. This should be contrasted with the study, already mentioned above, with regard to pore water[95] in which clays with various interlamellar spacings were studied. No dependence of the volume properties of the water with the cation exchange capacity, surface charge density or geometry was found. The dielectric relaxation times of water in various kaolinite clays[141] suggest that at very low coverages, the water molecules are very tightly bound ($\approx 10^{-4}$ s) whereas by the time the coverage reaches approximately a half monolayer, the relaxation times are similar to that of ice (2×10^{-6} s). The results for the sodium kaolinite (as compared to potassium and caesium) suggest that the smaller caption, being capable of stronger hydration, reduces the water mobility more.

7.2. *Zeolites*

Zeolites are aluminosilicate materials made of SiO_4 and AlO_4 tetrahedra with each oxygen being shared between two tetrahedra. They are closely akin to feldspars in structure; indeed the natural zeolites are derived by hydrothermal modification of feldspar-type minerals and accordingly they are frequently found in volcanically active areas. However, many artificial zeolites have been made in the laboratory and several are available industrially where they are used as molecular sieves, catalysts or catalytical supports. Zeolites are large framework structures enclosing cavities that contain large ions and/or water molecules which have sufficient mobility to allow ion-exchange and reversible dehydration. The number of cations associated with the zeolite is determined simply from electroneutrality considerations; for a zeolite containing only the singly charged cation M^+, the formula would be $M_n(AlO_2)_n(SiO_2)_m \cdot xH_2O$. Their technological importance relates to the different sized cavities that can capture different sized molecules and allow highly specific catalytic chemical reactions to occur; there is, in particular, great interest in the area of shape selective catalysis. According to one of the standard texts on the structure of geological materials, there are 22 different zeolites, although in several instances they differ only in replacement by different cations[142]. Of the artificial zeolites, the best known are the A, X and Y zeolites which have the formulae $M_{12}(AlO_2)_{12}(SiO2)_{12} \cdot xH_2O$, $M_{86}(AlO_2)_{86}(SiO_2)_{106} \cdot yH_2O$ and $M_{56}(Al_2)_{56}(SiO_2)_{136} \cdot zH_2O$. The X and Y zeolites in particular have extremely open structures with void volumes approaching 50%. Ion exchange is readily accomplished with most zeolites simply by treating the solid material as a slurry with an aqueous salt solution. Variation in cation can significantly alter the size of the pores. Thus, for example, the commonly available 3A, 4A and 5A molecular sieves are all type A zeolites with the cations respectively K_{12}, Na_{12} and $Ca_{4.5}Na_3$. It is thus clear that we may vary the cation composition of zeolites almost at will and, to a certain extent, vary the cavity sizes. In addition, the variation of cation will very much affect the hydration properties of the zeolite since a not insignificant part of the hydration water will be bound up in cation hydration. Thus the dielectric properties of water in various artificial zeolites[35] show considerable very low frequency components attributable to frustrated ionic conduction in the relatively large 'micropools' of water within the cavities and cages. Unlike clays and other layer structures, zeolite hydration water is lost continuously when they are heated rather than in separate stages at different temperatures; their adsorption isotherms are continuous rather than stepped, as with the clays. The cages and cavities of zeolites are highly rigid; indeed it is possible to completely dealuminate zeolites by vigorous reaction with concentrated acid under reflux and yet preserve the silicate skeleton[143]. As expected, the dealuminated zeolites have extremely open structures and adsorb gases very freely[144].

8. Concluding remarks

This review has endeavoured to review our current understanding of the hydration of inorganic surfaces. Considerable emphasis has been placed on the various experimental techniques available since it the author's opinion that a full understanding requires that research groups study the same sample (or else samples prepared under identical conditions and which give identical experimental results) with a battery of exeperimental approaches in parallel. The classical approaches, such as isotherms and calorimetry, will always hold a central position in such studies but a detailed molecular interpretation of the interactions occurring will almost certainly require diffraction/scattering techniques and/or spectroscopic measurements. Computer simulation is clearly still in its infancy; the work performed to date is still tentative and very fundamental questions relating to boundary conditions and the effects of a small sample are not yet resolved. However, the results already give us a tantalising glimpse of the type of information that could become available in the next few years, providing that supercomputing power continues to grow.

References

1. A. C. Zettlemoyer, F. J. Micale & K. Klier, in '*Water, A Comprehensive Treatise*', Vol. 5, (ed. F. Franks) Plenum Press, New York, 1978, chapter 5.
2. J. Clifford, in '*Water, A Comprehensive Treatise*' Vol. 5, (ed. F. Franks), Plenum Press, New York, 1978, chapter 2.
3. E. Forslind & A. Jacobsson, in '*Water, A Comprehensive Treatise*', Vol. 5 (ed. F. Franks) Plenum Press, New York, 1978, chapter 4.
4. S. Brunauer, P. H. Emmett & E. Teller, *J. Amer. Chem. Soc.* **60**, 309 (1938).

5. E. A. Guggenheim, '*Applications of Statistical Mechanics*', Clarendon Press, Oxford, 1966, chapter 11.
6. P. R. C. Gascoyne & R. Pethig, *J. Chem. Soc., Faraday Trans.* 1 **73**, 171 (1977).
7. R. D. Diehl & S. C. Fain, *Surface Sci.* **125**, 116. (1983).
8. S. Brunauer, '*The Adsorption of Gases and Vapors*', Princeton University Press, Princeton, New Jersey, 1945.
9. Ref. 1, p. 256 and references therein.
10. Ref. 1, p. 270.
11. A. C. Zettlemoyer, F. J. Micale & Y. K. Lui, *Ber. Bunsenges. Phys. Chem.* **71**, 286 (1967).
12. C. F. Jones, R. A. Reeve, R. Rigg, R. L. Segall, R. St. C. Smart & P. S. Turner, *J. Chem. Soc., Faraday Trans.* 1 **80**, 2617 (1984).
13. R. O. Watts & I. J. McGee, '*Liquid State Chemical Physics*', Wiley-Interscience, New York, 1976, chapter 3.
14. S. M. Thompson & K. E. Gubbins, *J. Chem. Phys.* **74**, 6467 (1981).
15. D. M. Heyes, M. Barber & J. H. R. Clarke, *J. Chem. Soc., Faraday Trans.* 2 **73**, 1485 (1977); **75**, 1469, 1485 (1979).
16. D. Nicholson & G. N. Parsonage, '*Computer Simulation and the Statistical Mechanics of Adsorption*', Academic Press, London, 1982.
17. S. M. Levine & S. H. Garofalini, *J. Chem. Phys.* **86**, 2997 (1987).
18. S. H. Garfalini & S. M. Levine, *J. Amer. Ceramic Soc.* **68**, 376 (1985).
19. N. I. Christou, J. S. Whitehouse, D. Nicholson & N. E. Parsonage, *Faraday Symp. Chem. Soc.* **16**, 139 (1981).
20. N. I. Christou, J. S. Whitehouse, D. Nicholson & N. E. Parsonage, *Mol. Phys.* **55**, 397 (1985).
21. B. Jonsson, *Chem. Phys. Lett.* **82**, 520 (1981).
22. R. Sonnenschein & K. Heinzinger, *Chem. Phys. Letts.* **102**, 550 (1983).
23. M. Marchesi, *Chem. Phys. Letts.* **97**, 224 (1983).
24. G. Barabino, C. Gavotti & M. Marchesi, *Chem. Phys. Letts.* **104**, 478 (1984).
25. C. Y. Lee, J. A. McCammon & P. J. Rossky, *J. Chem. Phys.* **80**, 4448 (1984).
26. J. P. Valeau & A. A. Gardner, *J. Chem. Phys.* **86**, 4162 (1987).
27. A. A. Gardner & J. P. Valeau, *J. Chem. Phys.* **86**, 4171 (1987).
28. N. Anastasiou, D. Fincham & K. Singer, *J. Chem. Soc., Faraday Trans.* 2 **79**, 1639 (1983).
29. D. J. Mulla, P. F. Low, J. H. Cushman & D. J. Diestler, *J. Colloid Interface Sci.*, **100**, 576 (1984).
30. E. Spohr & K. Heinzinger, *Chem. Phys. Letts.* **123**, 218 (1986).
31. N. G. Parsonage & D. Nicholson, *J. Chem. Soc., Faraday Trans.* 2 **83**, 663 (1987); **82**, 1521 (1986).
32. S. S. Dukhin, *Surface and Colloid Science*, Vol. 3 (ed. E. Matijevic) Wiley-Interscience, New York, 1969, pp. 83–165.
33. J. McConnell, '*Rotational Brownian Motion and Dielectric Theory*', Academic Press, London, 1980.
34. T. Ramdeen & L. A. Dissado, *J. Chem. Soc., Faraday Trans.* 1 **80**, 325 (1984).
35. A. R. Haidar & A. K. Jonscher, *J. Chem. Soc., Faraday Trans.* 1 **82**, 3535 (1986).

36. A. K. Jonscher & A. R. Haidar, *J. Chem. Soc., Faraday Trans.* 1 **82**, 3553 (1986).
37. H. Fellner-Feldegg, *J. Phys. Chem.* **73**, 616 (1969).
38. C. Bucci, R. Fieschi & G. Guidi, *Phys. Rev.* **148**, 816 (1966).
39. P. Pissis & D. Daoukaki-Diamanti, *Chem. Phys.* **101**, 95 (1986) and references therein.
40. C. H. Rochester & D.-A. Trebilco, *J. Chem. Soc., Faraday Trans.* 1 **74**, 1125 (1978).
41. D. M. Griffiths & C. H. Rochester, *J. Chem. Soc., Faraday Trans.* 1 **73**, 1510 (1977).
42. B. Liedberg, B. Ivarson, I. Lundstrom & W. R. Salaneck, *Progress in Colloid and Polymer Science*, (eds. H. G. Kilian & A. Weiss), Steinkopf Verlag; Darmstadt (FRG), **70**, 67, (1985).
43. G. E. Walrafren, in '*Water, A Comprehensive Treatise*', Vol. 1 (ed. F. Franks) Plenum Press, New York, 1972, p. 151.
44. W. A. P. Luck, in '*Water, A Comprehensive Treatise*', Vol 2 (ed. F. Franks) Plenum Press, New York, 1973, p. 235.
45. P. G. Hall, A. Pidduck & C. J. Writht, *J. Colloid Interface Sci.* **79**, 339 (1981).
46. P. B. Barraclough & P. G. Hall, *J. Chem. Soc., Faraday Trans.* 1 **74**, 1360 (1978).
47. J. W. Clark & P. G. Hall, *J. Chem. Soc., Faraday Trans.* 1 **81**, 2067 (1985).
48. D. M. Kroll & J. G. van Lierop, *J. Non-Crystalline Solids* **68**, 163 (1984).
49. T. P. Mernagh & R. P. Cooney, *J. Chem. Soc., Faraday Trans.* 1 **80**, 3469 (1984).
50. H. P. Kung & T. T. Chen, *Chem. Phys. Letts.* **130**, 311 (1986).
51. D. Roy & T. E. Furtak, *Chem. Phys. Letts.* **129**, 501 (1986).
52. D. Roy & T. E. Furtak, *Chem. Phys. Letts.* **124**, 299 (1986).
53. S. Coluccia, A. J. Tench & R. L. Segall, *J. Chem. Soc., Faraday Trans.* 1 **75**, 1769 (1979).
54. S. Coluccia, A. M. Deane & A. J. Tench, *J. Chem. Soc., Faraday Trans.* 1 **74**, 2913 (1978).
55. M. Breysse, B. Claudel, L. Faure & M. Guenin, *J. Colloid Interface Sci* **70**, 201 (1979).
56. N. Lahav, L. Coyne & J. G. Lawless, *Clays and Clays Minerals* **33**, 207 (1985).
57. W. Derbyshire, '*Water, A Comprehensive Treatise*' Vol. 7 (ed. F. Franks) Plenum Press; New York 1982.
58. A. K. Covington & K. E. Newman, '*Modern Aspects of Electrochemistry*, No. 12' (ed. J. O'M. Bockris & B. E. Conway) Plenum Press, New York, 1977.
59. G. K. Rennie & J. Clifford, *J. Chem. Soc., Faraday Trans.* 1 **73**, 680 (1977).
60. J. Tabony, G. Bomchil, N. M. Harris, M. Leslie, J. W. White, P. H. Gamlen, R. K. Thomas & T. D. Trewern, *J. Chem. Soc., Faraday Trans.* 1 **75**, 1570 (1979).
61. A. Abragam, '*Principles of Magnetic Resonance*'. Oxford University Press, Oxford, 1961.
62. T. C. Farrar & E. D. Becker, '*Pulse and Fourier Transform NMR*'. Academic Press, New York, 1971.

63. M. Kadif-Hanifi, *Clays and Clay Minerals* **28**, 65 (1980).
64. J. Hougardy, W. E. E. Stone & J. J. Fripiat, *J. Chem. Phys.* **64**, 3840 (1976).
65. G. Martini, *Colloids and Surfactants* **11**, 409, (1984).
66. G. Martini, F. Ottaviani & M. Romanelli, *J. Colloid Interface Sci.* **84**, 105 (1983).
67. V. Bassetti, L. Burlamacchi & G. Martini, *J. Amer. Chem. Soc.* **101**, 5471 (1979).
68. G. Martini, *J. Colloid Interface Sci.* **80**, 39 (1981).
69. P. Brüggeller, *J. Colloid Interface Sci.* **94**, 524 (1983).
70. J. E. Enderby & G. W. Neilson, *Rep. Prog. Phys.* **44**, 593–653 (1981).
71. J. E. Enderby, *Ann. Rev. Phys. Chem.* **34**, 155 (1983).
72. P. Meehan, T. Rayment, R. K. Thomas, G. Bomchil & J. W. White, *J. Chem. Soc., Faraday Trans.* 1 **76**, 2011 (1980).
73. D. C. Steytter, J. C. Dore & C. J. Wright, *Mol. Phys.* **48**, 1031 (1983).
74. J.-J. Legendre & J. Livage, *J. Colloid Interface Sci.*, **94**, 75 (1983).
75. J.-J. Legendre, P. Alderbert, N. Baffier & J. Livage, *J. Colloid Interface Sci.* **94**, 84 (1983).
76. P. Alderbert, H. W. Haesslin, N. Baffier & J. Livage, *J. Colloid Interface Sci.* **98**, 478 (1984).
77. P. Alderbert, H. W. Haesslin, N. Baffier & J. Livage, *J. Colloid Interface Sci.* **98**, 484 (1984).
78. P. J. Quinn & L. J. Lis, *J. Colloid Interface Sci.* **115**, 220 (1987).
79. D. J. Cebula, R. K. Thomas & J. W. White, *J. Chem. Soc., Faraday Trans.* 1 **76**, 314 (1980).
80. J. T. Tuck, P. L. Hall & M. H. B. Hayes, *J. Chem. Soc., Faraday Trans.* 1 **80**, 309 (1984).
81. G. Bomchil, N. Harris, M. Leslie, J. Tabony, J. W. White, P. H. Gamlen, R. K. Thomas & T. D. Trewern, *J. Chem. Soc., Faraday Trans.* 1 **75**, 1535 (1979).
82. P. H. Gamlen, R. K. Thomas, T. D. Trewern, G. Bomchil, N. M. Harris, M. Leslie, J. Tabony & J. W. White, *J. Chem. Soc. Faraday Trans.* 1 **75**, 1542 (1979).
83. *Idem, J. Chem. Soc., Faraday Trans.* 1 **75**, 1534 (1979).
84. J. Tabony, G. Bomchil, N. M. Harris, M. Leslie, J. W. White, P. W. Gamlen, R. K. Thomas & T. E. Trewern, *J. Chem. Soc., Faraday Trans.* 1 **75**, 1570 (1979).
85. J. Schneir, R. Sonnenfeld, P. K. Hansma & J. Tersoff, *Phys. Rev. B* **34**, 4979 (1986).
86. D. M. Hercules & S. H. Hercules, *J. Chem. Educ.* **61**, 402 (1984).
87. D. M. Hercules & S. H. Hercules, *J. Chem. Educ.* **61**, 483 (1984).
88. D. M. Hercules & S. H. Hercules, *J. Chem. Educ.* **61**, 592 (1984).
89. R. B. Stewart, *Canadian Chemical News* **11** (1984).
90. F. Franks, '*Polywater*', MIT Press, Cambridge, USA, 1981.
91. W. Drost-Hansen, in '*Biophysics of Water*' (ed. F. Franks) Wiley, New York, 1982.
92. F. M. Etzler & D. M. Fagundus, *J. Colloid Interface Sci.* **115**, 513 (1987).
93. Y. Sun, H. Lin & P. F. Low, *J. Colloid Interface Sci.* **112**, 556 (1986).
94. B. V. Derjaguin, V. V. Karasev & E. N. Khromova, *J. Colloid Interface Sci.* **109**, 586 (1986).

95. B. V. Derjaguin, V. V. Karasev & E. N. Khromova, *J. Colloid Interface Sci.* **78**, 273 (1980).
96. F. M. Etzler & D. M. Fagandus, *J. Colloid Interface Sci.* **92**, 43 (1983).
97. D. J. Mulla & F. P. Low, *J. Colloid Interface Sci.* **95**, 51 (1983).
98. M. E. Schrader, *J. Colloid Interface Sci.* **100**, 372 (1984).
99. S. K. Hara & N. C. Debnath, *Chem. Phys. Letts.* **121**, 490 (1985).
100. K. Takahashi, *J. Chem. Soc., Faraday Trans.* 1 **78**, 2059 (1982).
101. K. Takahashi, *J. Colloid Interface Sci.* **88**, 286 (1983).
102. W. Smit, *J. Colloid Interface Sci* **84**, 272 (1982).
103. L. Nondek, *J. Chromatog.* **238**, 264 (1982).
104. P. B. Barraclough & P. G. Hall, *J. Chem. Soc., Faraday Trans.* 1 **74**, 1360 (1977).
105. C. Poinsignon & J. D. F. Ramsay, *J. Chem. Soc., Faraday Trans.* 1 **82**, 3447 (1986).
106. J. Bunce, J. D. F. Ramsay & J. Penfold, *J. Chem. Soc., Faraday Trans.* 1 **81**, 2845 (1985).
107. J. Penfold & J. F. D. Ramsay, *J. Chem. Soc., Faraday Trans.* 1 **81**, 117 (1985).
108. D. C. Steytler & J. C. Dore, *Mol. Phys.* **56**, 1001 (1985).
109. P. A. Sermon, *J. Chem. Soc., Faraday Trans.* 1 **76**, 885 (1980).
110. L. Piculell & B. Halle, *J. Chem. Soc., Faraday Trans.* 1 **82**, 387, 401, 415 (1986).
111. P. G. Hall, R. T. Williams & R. C. T. Slade, *J. Chem. Soc., Faraday Trans.* 1 **81**, 847 (1985).
112. P. G. Hall, A. Pidduck & C. J. Wright, *J. Colloid Interface Sci*, **79**, 339 (1981).
113. C. H. Rochester & S. A. Topham, *J. Chem. Soc., Faraday Trans.* 1 **75**, 1073 (1979).
114. C. H. Rochester & S. A. Topham, *J. Chem. Soc., Faraday Trans.* 1 **75**, 591 (1979).
115. T. Ishikawa, S. Nitta & S. Kondo, *J. Chem. Soc., Faraday Trans.* 1 **82**, 2401 (1986).
116. D. E. Yates, R. O. James & T. W. Healy, *J. Chem. Soc., Faraday Trans.* 1 **76**, 1 (1980).
117. D. M. Griffiths & C. H. Rochester, *J. Chem. Soc., Faraday Trans.* 1 **73**, 1510 (1977).
118. T. M. El-Akkad, *J. Colloid Interface Sci.* **76**, 67 (1980).
119. T. Miyazaki, Y. Kuroda, K. Morishige, S. Kittaka, J. Umemura, T. Takenaka & T. Morimoto, *J. Colloid Interface Sci.* **106**, 154 (1985).
120. D. A. Griffiths & D. W. Fuerstenau, *J. Colloid Interface Sci.* **80**, 271 (1981).
121. R. R. Bailey & J. P. Wrightman, *J. Colloid Interface Sci.* **70**, 112 (1979).
122. J. L. G. Fierro, S. Mendioroz & J. Sanz, *J. Colloid Interface Sci.* **93**, 487 (1983).
123. S. Kittaka, K. Morishiga, J. Nishiyama & T. Morimoto, *J. Colloid Interface Sci.* **91**, 117 (1983).
124. R. B. Fahem, M. I. Zaki & N. H. Yacoub, *J. Colloid Interface Sci.* **88**, 502 (1983).
125. Y. Kawai, M. Nitta & K. Aomura, *J. Colloid Interface Sci.* **88**, 47 (1983).

126. C. L. Cronan, F. J. Micale, A. C. Zettlemoyer, M. Topic & H. Leidheiser, Jr, *J. Colloid Interface Sci*, **75**, 43 (1980).
127. K. Morishige, S. Kittaka & T. Moriyasu, *J. Chem. Soc., Faraday Trans.* 1 **76**, 728 (1980).
128. S. Kittaka, S. Kanemoto & T. Morimoto, *J. Chem. Soc., Faraday Trans.* 1 **74**, 676 (1978).
129. J. S. Chung, R. Miranda & C. O. Bennett, *J. Chem. Soc., Faraday Trans.* 1 **81**, 19 (1985).
130. (a) A. E. Regazzoni, M. A. Blesa & J. G. Maroto, *J. Colloid Interface Sci.* **91**, 560 (1983); (b) F. G. R. Gimblett, A. A. Rahman & K. S. W. Sing, *J. Colloid Interface Sci.*, **84**, 337 (1982).
131. M. Breysse, B. Claudel, L. Faure & M. Guenin, *J. Colloid Interface Sci.* **70**, 201 (1979).
132. J. L. Stakebake & H. N. Robinson, *J. Colloid Interface Sci.* **95**, 37 (1983).
133. R. Gobinathan, K. Hariharan & P. Ramasamy, *J. Colloid Interface Sci.* **86**, 284 (1983).
134. P. B. Barraclough & P. G. Hall, *J. Chem. Soc., Faraday Trans.* 1 **72**, 610 (1976).
135. P. B. Barraclough & P. G. Hall, *J. Chem. Soc., Faraday Trans.* 1 **71**, 2266 (1975).
136. P. B. Barraclough & P. G. Hall, *Surface Sci.* **46**, 393 (1975).
137. M. C. Ball & L. S. Norwood, *J. Chem. Soc., Faraday Trans.* 1 **74**, 1477 (1978).
138. L. Pauling, *Proc. Natl. Acad. Sci. U.S.* **16**, 123 (1930).
139. R. E. Grim, '*Clay Minerology*', McGraw Hill, New York, 1968.
140. G. W. Brindley & G. Brown, '*Crystal Structures of Clay Minerals and their X-Ray Identification*', Minerological Society, London, 1980.
141. P. G. Hall & M. A. Rose, *J. Chem. Soc., Faraday Trans.* 1 **74**, 1221 (1978).
142. W. A. Deer, R. A. Howie & J. Zussman, '*Rock Forming Minerals*, Vol. 4, Framework Silicates', Longmans, London, 1963.
143. R. M. Barrer & J.-C. Trombe, *J. Chem. Soc., Faraday Trans.* 1 **74**, 2786 (1978).
144. R. M. Barrer & J.-C. Trombe, *J. Chem. Soc., Faraday Trans.* 1 **74**, 2798 (1978).